World Scientific Series in Current Energy Issues   Volume 3

Energy from
the Nucleus

The Science and Engineering of Fission and Fusion

World Scientific Series in Current Energy Issues

Series Editor: Gerard M Crawley *(University of South Carolina & Marcus Enterprise LLC, USA)*

World Scientific Series in Current Energy Issues   Volume 3

Energy from the Nucleus

The Science and Engineering of Fission and Fusion

Editor

Gerard M Crawley

Marcus Enterprise LLC, USA

&

Professor and Dean Emeritus
University of South Carolina, USA

 World Scientific

NEW JERSEY · LONDON · SINGAPORE · BEIJING · SHANGHAI · HONG KONG · TAIPEI · CHENNAI · TOKYO

Published by

World Scientific Publishing Co. Pte. Ltd.

5 Toh Tuck Link, Singapore 596224

USA office: 27 Warren Street, Suite 401-402, Hackensack, NJ 07601

UK office: 57 Shelton Street, Covent Garden, London WC2H 9HE

Library of Congress Control Number: 2015048898

British Library Cataloguing-in-Publication Data
A catalogue record for this book is available from the British Library.

World Scientific Series in Current Energy Issues — Vol. 3
ENERGY FROM THE NUCLEUS
The Science and Engineering of Fission and Fusion

ISBN 978-981-4689-19-9

Desk Editors: Herbert Moses/Amanda Yun

Typeset by Stallion Press
Email: enquiries@stallionpress.com

Printed in Singapore

Foreword to the World Scientific Series on Current Energy Issues

Sometime between 400,000 and a million years ago, an early humanoid species developed the mastery of fire and changed the course of our planet. Still, as recently as a few hundred years ago, the energy sources available to the human race remained surprisingly limited. In fact, until the early 19th century, the main energy sources for humanity were biomass (from crops and trees), their domesticated animals, and their own efforts.

Even after many millennia, the average per capita energy use in 1830 only reached about 20 Gigajoules (GJ) per year. By 2010, however, this number had increased dramatically to 80 GJ per year.[1] One reason for this notable shift in energy use is that the number of possible energy sources increased substantially during this period, starting with coal in about the 1850s and then successively adding oil and natural gas. By the middle of the 20th century, hydropower and nuclear fission were added to the mix. As we move into the 21st century, there has been a steady increase in other forms of energy such as wind and solar, although presently they represent a relatively small fraction of world energy use.

Despite the rise of a variety of energy sources, per capita energy use is not uniform around the world. There are enormous differences from country to country, pointing to a large disparity in wealth and opportunity, see Table 1. For example, in the United States the per capita energy use per year in 2011 was 312.8 million Btu[a] (MMBtu) and in Germany, 165.4 MMBtu. In China, however, per capita energy use was only 77.5 MMBtu, despite its impressive economic and technological gains. India, weighs in even lower at 19.7 MMBtu per person.[2] The general trends over the last decade suggest that countries with developed economies generally show

---

[a] *Note*: 1 GJ = 0.947 MMBtu.

Table 1:   Primary Energy Use per Capita in Million Btu (MMBtu).[2]

Country	2007 (MMBtu)	2011 (MMBtu)	Percentage change
Canada	416.1	393.7	−5.4
United States	336.9	312.8	−7.2
Brazil	52.7	60.2	14.2
France	175.7	165.9	−5.6
Germany	167.8	165.4	−1.4
Russia	204.0	213.4	4.6
Nigeria	6.1	5.0	−18.0
Egypt	36.4	41.6	14.3
China	57.1	77.5	35.7
India	17.0	19.7	15.9
World	**72.2**	**74.9**	**3.7**

modest increases or even small decreases in energy use, but that developing economies, particularly China and India are experiencing rapidly increasing energy consumption per capita.

These changes, both in the kind of resource used and the growth of energy use in countries with developing economies, will have enormous effects in the near future, both economically and politically, as greater numbers of people compete for limited energy resources at a viable price. A growing demand for energy will have an impact on the distribution of other limited resources such as food and fresh water. All these lead to the conclusion that energy will be a pressing issue for the future of humanity.

Another important consideration is that all energy sources have disadvantages as well as advantages, risks as well as opportunities, both in the production of the resource and in its distribution and ultimate use. Coal, the oldest of the "new" energy sources, is still used extensively to produce electricity, despite its potential environmental and safety concerns in mining both underground and open cut mining. Burning coal releases sulfur and nitrogen oxides which in turn can lead to acid rain and a cascade of detrimental consequences. Coal production requires careful regulation and oversight to allow it to be used safely and without damaging the environment. Even a resource like wind energy using large wind turbines has its critics because of the potential for bird kill and noise pollution. Some critics also find large wind turbines an unsightly addition to the landscape, particularly when the wind farms are erected in pristine environments. Energy from nuclear fission, originally believed to be "too cheap to meter"[3] has not had the growth predicted because of the problem with long-term storage of the waste from nuclear reactors and because of the public perception regarding the danger of catastrophic accidents such as happened at Chernobyl in 1986 and at Fukushima in 2011.

Even more recently, the measured amount of CO_2, a greenhouse gas, in the global atmosphere has steadily increased and is now greater than 400 parts per million (ppm).[4] This has raised concern in the scientific community and has led the majority of climate scientists to conclude[5] that this increase in CO_2 will produce an increase in global temperatures. We will see a rise in ocean temperature, acidity, and sea level, all of which will have a profound impact on human life and ecosystems around the world. Relying primarily on fossil fuels far into the future may therefore prove precarious, since burning coal, oil, and natural gas will necessarily increase CO_2 levels. Certainly for the long term future, adopting a variety of alternative energy sources which do not produce CO_2 seems to be our best strategy.

The volumes in the *World Scientific Series on Current Energy Issues* explore different energy resources and issues related to the use of energy. The volumes are intended to be comprehensive, accurate, current, and international perspective. The authors of the various chapters are experts in their respective fields and provide reliable information that can be useful not only to scientists and engineers, but also to policy makers and the general public interested in learning about the essential concepts related to energy. The volumes will deal with the technical aspects of energy questions but will also include relevant discussion about economic and policy matters. The goal of the series is not polemical but rather is intended to provide information that will allow the reader to reach conclusions based on sound, scientific data.

The role of energy in our future is critical and will become increasingly urgent as world population increases and the global demand for energy turns ever upwards. Questions such as which energy sources to develop, how to store energy, and how to manage the environmental impact of energy use will take center stage in our future. The distribution and cost of energy will have powerful political and economic consequences and must also be addressed. How the world deals with these questions will make a crucial difference to the future of the earth and its inhabitants. Careful consideration of our energy use today will have lasting effects for tomorrow. We intend that the *World Scientific Series on Current Energy Issues* will make a valuable contribution to this discussion.

References

1. Our Finite World: World energy consumption since 1820 in charts. Accessed in February 2015 at http://ourfiniteworld.com/2012/03/12/world-energy-consumption-since-1820-in-charts/.
2. U.S. Energy Information Administration, Independent Statistics & Analysis. Accessed in March 2015 at http://www.eia.gov/cfapps/ipdbproject/iedindex3. cfm?tid=44&pid=45&aid=2&cid=regions&syid=2005&eyid=2011&unit=MB TUPP.
3. The quote is from a speech by Lewis Strauss, then Chairman of the United States Atomic Energy Commission, in 1954. There is some debate as to whether Strauss actually meant energy from nuclear fission or not.
4. NOAA Earth System Research Laboratory, Trends in Atmospheric Carbon Dioxide. Accessed in March 2015 at http://www.esrl.noaa.gov/gmd/ccgg/ trends/.
5. IPCC, Intergovernmental Panel on Climate Change, Fifth Assessment report 2014. Accessed in March 2015 at http://www.ipcc.ch/.

Introduction to Energy from the Nucleus

The first few decades of the 20th century saw the rapid increase of our understanding of the nature of the physical world. We began to understand the structure of the atom with its tiny, heavy nucleus surrounded by a cloud of electrons and even showed that the nucleus consisted of protons and neutrons. We also learned that mass is a form of energy and can be converted to more familiar forms through the famous equation:

$$E = mc^2,$$

where c is the speed of light and is a very large number viz. 3×10^8 m/sec. Because c^2 (about 9×10^{16} m^2/sec^2) is so large this means that a small amount of mass can be converted into a large amount of energy.

There are two rather different processes that occur in the nuclear realm which result in the conversion of mass into large amounts of energy. One process is **FISSION** where a large nucleus breaks apart into smaller pieces releasing energy. The other is **FUSION** where light nuclei combine into a heavier one which also releases energy. This latter process is the one that powers the stars. Both of these processes are discussed in much more detail in this volume on *Energy from the Nucleus*.

The first application of both of these processes was military. A bomb, based on fission (the "atomic" bomb) was developed during World War II (1939–1945) and was used on two Japanese cities, Hiroshima and Nagasaki. After the war, an even more powerful bomb was developed based on the fusion process (the "hydrogen" bomb) and fortunately it has never been used but only tested.

While the first use of nuclear energy as an engine of destruction has given a taint to the concept, the production of energy from both fission and fusion for civilian use was attempted as soon as the war was over. The fission process was the easier one to realize in practice and the first nuclear reactors based on fission were in operation during the 1950s.

The present status of the current and future generations of nuclear fission reactors is discussed in Chapters 1–3, including new safety requirements and the possible environmental impact.

There are two general approaches to producing a controlled nuclear fusion reactor namely, inertial confinement fusion and magnetic confinement fusion. In inertial confinement, beams typically of laser light impinge on a target and attempt to heat it sufficiently that a nuclear fusion reaction is initiated. In magnetic confinement, a plasma of positive ions is confined in a shaped magnetic field and heated to attempt to produce the conditions for fusion.

While we know that nuclear fusion works (witness the sun), but in spite of the early optimism, it has proved elusive to produce a controlled fusion reactor on earth. Chapters 4 and 5 discuss two different methods of attempting inertial confinement fusion, direct and indirect. While both approaches have shown promise and many advances have been made, producing more energy out from the reaction than has to be input, has proved extremely difficult.

Chapter 6 examines in detail the alternative approach of magnetic confinement. The different possible magnetic field conditions and achievements are discussed together with possible future directions. The final chapter presents the status of the large international project ITER being constructed in the south of France to test many of the ideas both for confining the plasma and for heating it. This is an important step on the road to achieve a working commercial fusion energy reactor.

Both nuclear fission reactors and a potential fusion reactor have numerous advantages as well as some potential disadvantages. For fission reactors, the technology is quite mature although the industry still needs to continue to improve the technology especially the safety aspects. Fission reactors require much less amounts of uranium than a coal fired plant does coal with the attendant decrease in transportation infrastructure and dangers. If the breeder reactor is deployed, nuclear fission will be able to provide electrical power for many thousands of years. Plus there is neither carbon dioxide (CO_2) or methane emission from a nuclear fission plant, so that it does not contribute to the greenhouse effect and global warming. On the other hand, the consequences of a reactor accident and especially a meltdown are so great that safety concerns must remain a high priority. Another disadvantage of fission reactors is the need to store highly radioactive residues for many centuries.

A fusion reactor would have enormous advantages. It is intrinsically safer with no chance of a runaway reaction. Fusion reactors use the heavy isotopes of hydrogen as fuel which is an essentially unlimited resource. There are again no CO_2 emissions from a fusion reactor, and little or no long lived radioactive residues that need to be stored. The remaining issue, and this is no easy one to predict, is whether fusion plants can be built to operate reliably and economically. The potential payoffs in clean, safe power are so great that the risk of not making the investment outweighs the risk of failure. The international collaboration on ITER is a giant step in the right direction.

Gerard M. Crawley

Contents

Chapter 4: Indirect-Drive Inertial Confinement Fusion   69

Erik Storm and John D. Lindl

Chapter 6: Magnetic Fusion Energy 165

M. C. Zarnstorff and R. J. Goldston

Chapter 7: Creating A Star — The Global ITER Partnership   189

M. Uhran

Chapter 1

Fundamentals of Nuclear Fission

Bertrand Barré

Institut National des Sciences et Techniques Nucléaires
CEA Saclay, 91191 Gif sur Yvette Cedex, France
bcbarre@wanadoo.fr

When a few heavy nuclei, called fissile nuclei, absorb a neutron, they split into two fragments, releasing a huge amount of heat in the process. This heat is used in nuclear plants to generate electric power and feed it to the grid. Fission occurs in the core of the reactor which constitutes the nuclear part of the plant. The core is made of fuel assemblies where the fission chain reaction is produced and controlled, and the radioactive fission products are contained to protect human health and environment from the release of radioactivity. The fuel cycle is the series of industrial steps from the uranium ore to the management of the spent fuel assemblies.

1 Introduction

Democritus, (460–370 BC) as early as 400 BC, formulated the hypothesis that matter was made of atoms assembled together and surrounded by a void. However, Aristotle (364–322 BC) opposed this view because he was convinced that Nature "abhorred" a vacuum, and his influence was so dominant that the atomistic theory was ignored for centuries. It was only revived by the work of Lavoisier (1743–1794) and Dalton (1766–1844) who transformed Alchemy into Chemistry. But the nature of the atom itself remained a mystery until the very beginning of the 20th century.

The turning point was the discovery of radioactivity by Becquerel and Pierre and Marie Curie in 1896. Suddenly scientific discoveries due to Einstein, Rutherford, de Broglie, Bohr, Chadwick, Heisenberg, and Planck, to name only a few, bloomed, thus creating a revolution in Physics. By 1932, the model of the atom described below was agreed to among the scientists. "Artificial" radioactivity was discovered by Irene and Frederic Joliot-Curie in 1934 and experiments on neutron capture by Fermi led to the discovery of fission by Hahn, Strassmann, and Meitner in 1938 and of the chain reaction by the Joliot team in 1939.

Unfortunately, due to World War II, the first application of this newly discovered energy source was the atomic bomb, developed within the Manhattan Project. Only in 1954 was a nuclear power plant connected to the grid, in Obninsk (Soviet Union).

2 Radioactivity, Fission, Fusion

Whether solid, liquid, or gaseous matter is made of atoms, alone or combined to form molecules or crystals. Nearly all the mass of the atom is concentrated in a positively charged nucleus, which is surrounded by "clouds" of negatively charged electrons. The nucleus itself is an assembly of positively charged protons plus neutrons which carry no electrical charge. Neutrons and protons have essentially the same mass, and their number A determines the atomic mass of a given atom. The number of protons, Z, identical to the number of electrons in an electrically neutral atom, determines the chemical nature of a given element, from hydrogen $(Z = 1)$ to uranium $(Z = 92)$ and beyond. Nuclides having the same Z but different A are *isotopes* of the same chemical element. There is a relationship between A and Z, which determines the stability of a nucleus. Stable elements are found in the so-called valley of stability in an (A, Z) diagram.

All nuclides found in nature are not stable: Some do not have the right balance of neutrons and protons in their nucleus and, over time, they tend to reach the valley of stability by emitting radiation. This is called *radioactive decay* or radioactivity. For instance, in a nucleus with too many neutrons, one neutron will turn into a proton and one electron will be expelled from the nucleus, which keeps the atom neutral. This is called β^- decay. The new nuclide, whose charge is now Z^{+1}, belongs to a different element. There is also the possibility of a neutron decaying to a proton within the nucleus and emitting a positive electron or positron (β^+). In this case, the new nucleus has a charge of Z^{-1}. If the nucleus has too many neutrons and protons, it will expel an assembly made of two neutrons and two protons called an α particle, identical to the nucleus of a helium atom. If the nucleus has too much energy, it gets rid of its excess energy by emitting electromagnetic radiation, a γ ray.

If a nucleus absorbs a particle, it will become unstable and undergo radioactive decay. The new nuclide will be "artificial", but the radiation it emits will still be "natural", α, β, or γ.

When a few heavy nuclei (^{233}U, ^{235}U, ^{239}Pu, ^{241}Pu) absorb a neutron, they become so unstable that they split into two **fragments**, releasing an enormous amount of energy and emitting two or three new neutrons with high velocity, around 20,000 km/sec (Fig. 1).

One such **fission** releases about 200 million electron volts (MeV) (Table 1), while a typical chemical reaction releases only a few electron volts (eV). The energy of the neutrinos emitted during the β^- decay cannot be recovered because neutrinos interact so weakly with matter.

The new neutrons released during fission can, in turn, be absorbed by another **fissile** nucleus and propagate a **chain reaction** of successive fissions. The probability of a neutron being absorbed increases when the velocity of the neutron decreases to the average velocity of the surrounding nuclei, due to thermal agitation. The velocity of those "slow" or "thermal" neutrons is close to 2.2 km/sec.

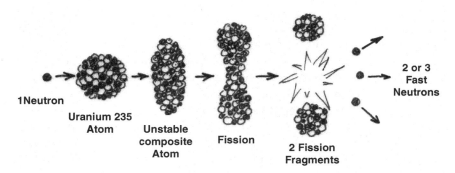

Fig. 1. Fission (schematic).

Table 1: Energy Released in the Fission of One ^{235}U Nucleus by a Thermal Neutron.[1]

Fission fragments kinetic energy	166 MeV
Fission products β^- decay	7 MeV
Fission products γ decay	7 MeV
Prompt neutrons	5 MeV
Prompt γ	8 MeV
(neutrinos)	(10 MeV)
Total recoverable energy	193 MeV

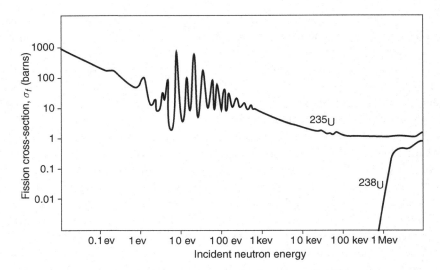

Fig. 2.   Neutron cross-sections for uranium to fission.[2]

Cross-sections (denoted σ) are a measure of the target surface that a nucleus offers to a neutron for a given nuclear reaction, i.e. a measure of the probability that such a reaction occurs. They are expressed in ***barns*** (1 barn $= 10^{-24}$ cm^2). Cross-sections are specific to a particular interaction between a given nucleus with a neutron, and they are highly dependent upon the velocity of the neutron.

As shown in Figs. 2 and 3, there are three regions in the cross-section curves: At low neutron energy, σ is proportional to the inverse of the neutron velocity. Then there is a ***resonance*** region where σ may vary greatly for a small energy difference. At high energy, above 1 MeV, some ***threshold*** reactions may suddenly appear like the "fast" fission of ^{238}U. Detailed cross-section plots are available from the Evaluated Nuclear Data File at Brookhaven National Laboratory.[2] Table 2 summarizes the main data useful in reactor design.

The fission of a nucleus of ^{235}U emits on average 2.4 ***prompt*** neutrons, and the figure for neutrons per fission (ν) reaches 2.9 for plutonium. But a few additional neutrons — their percentage is denoted $\beta_{\text{eff}} = 0.2$–0.6% — are emitted seconds or minutes later, during the radioactive decay of some fission fragments.

These ***delayed*** neutrons are essential to control and stabilize the chain reaction in the core of a nuclear reactor: if there were only prompt neutrons,

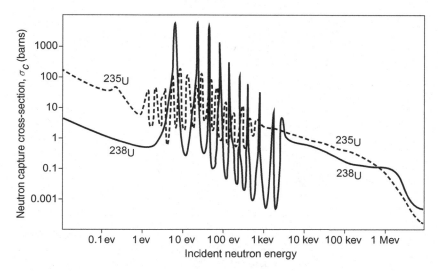

Fig. 3.   Cross-sections for uranium to capture a neutron (absorption without fission).[2]

Table 2:   Main Neutron Data for Fissile and Fertile Nuclei (Rounded Figures).

		^{233}U	^{235}U	^{239}Pu	^{232}Th	^{238}U
F	Fission σ_f barns	2.8	2.0	1.9	0.01	0.05
	Capture σ_c barns	0.3	0.5	0.6	0.35	0.30
	Neutrons/fission, ν	2.5	2.5	2.9	2.3	2.75
	Neutrons/absorption, η	2.2	2.0	2.2	—	—
T	σ_f barns	527	579	741	—	2.7
	σ_c barns	54	100	265	7.6	—
	ν	2.5	2.4	2.9	—	—
	η	2.2	2.0	2.1	—	—
	Delayed neutrons, (β_{eff}) %	0.28	0.64	0.21	2.3	1.5

Note: F = Fast neutrons, T = Thermal neutrons.

one could build nuclear bombs but one could not use nuclear power. The number $\eta = \nu \times \sigma_f/(\sigma_f + \sigma_c)$ is the number of neutrons emitted per neutron absorbed by a fissile nucleus. For breeding purposes, η must be significantly larger than 2.0.

A few heavy nuclides (^{232}Th, ^{238}U) are called *fertile*. When they absorb a neutron, they undergo two successive β-decays and turn into a *fissile* nuclide, respectively ^{233}U and ^{239}Pu. The uranium found in the

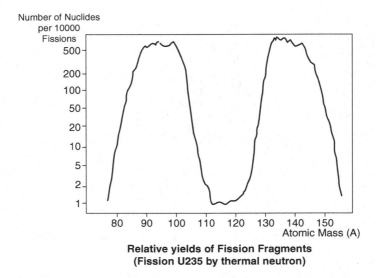

Relative yields of Fission Fragments
(Fission U235 by thermal neutron)

Fig. 4.   Distribution of fission products according to their atomic mass.

Earth's crust is mostly made of 99.3% fertile ^{238}U and 0.7% fissile ^{235}U, while thorium is 100% ^{232}Th and has no fissile isotope.

Fission is a complex phenomenon that gives birth to a wide variety of pairs of fission fragments, the masses of which are distributed as shown in Fig. 4. Most of those fragments are radioactive and decay into "daughter" nuclides. The fragments and their daughters are called *fission products*.

Each radioactive nucleus is characterized by its *half-life*, denoted $T^{1/2}$, which measures the time necessary for half of a given number of such nuclei to have decayed. Even though a *single* "short-lived" nucleus may survive almost indefinitely, statistics allows us to measure half-lives with incredible precision. Table 3 gives the value of the half-life of a number of nuclei of interest, and Table 4 gives the activities of some fission products as they decay with time. Activities in this table are expressed in Watts per kg of product.

3 How Does A Nuclear Reactor Operate?

A nuclear reactor, or nuclear power plant, is a machine within which a fission chain reaction is maintained to generate electricity. The chain reaction occurs in the *core* of the reactor, made from a number of *fuel assemblies* (or subassemblies) which contain the nuclear fuel. The heat released by

Table 3: Half-Lives of Some Natural (N) and Artificial (A) Nuclides.

Nuclide	N/A	Half-life	Uses
Tritium ^{3}H	A	12.3 years	Nuclear fusion
Carbon 14	N	5730 years	Dating old artifacts
Oxygen 15	A	2.02 min	Medical imaging
Cobalt 60	A	5.27 years	Industrial gammagraphy
Cesium 137	A	30.2 years	Cancer therapy
Radon 222	N	3.82 days	Source of natural irradiation
Radium 226	N	1,600 years	Used previously, now banned
Uranium 235	N	704 million years	Nuclear fission
Uranium 238	N	4.54 billion years	Nuclear fission
Plutonium 239	A	24,100 years	Nuclear fission

Table 4: Typical Fission Product Activities in 1 kg of Total Fission Products.[3]

Nuclide	Half-life	Activity (W) 1 Year	Activity (W) 10 Years	After 100 Years	After 1000 Years
^{143}Ba	12 sec				
^{131}I	8 days				
^{103}Ru	40 days	20			
^{144}Ce	284 days	2.5×10^4	9		
^{106}Ru	1 year	1.9×10^3	2		
^{85}Kr	11 years	3×10^2	2×10^2	0.6	
^{90}Sr	29 years	3×10^3	2.5×10^3	2.9×10^2	
^{137}Cs	30 years	3.2×10^3	2.6×10^3	3.2×10^2	
^{151}Sm	90 years	4	4	2	0.002
^{99}Tc	200,000 years	0.5	0.5	0.5	0.5
^{129}I	1.7 million years	0.001	0.001	0.001	0.001

fission is carried off by a *coolant* (gaseous or liquid) and used to boil water and produce high pressure steam. This steam activates a turbine coupled to an electrical generator.

In a pressurized water reactor (PWR), the type most used in the world at present, the coolant is ordinary water maintained at high pressure, viz. 15.5 MPa, so as to remain liquid at temperatures above 300°C. This water circulates within a closed *primary circuit* which includes the core, located in a *pressure vessel* made of thick high strength steel, and several *loops*, 2–4 according to the size of the plant. Each loop comprises a *primary pump* and a *steam generator*. One of the loops is connected to a *pressurizer* in which a steam bubble maintains the required pressure within the primary circuit (Fig. 5).

PWR Schematic

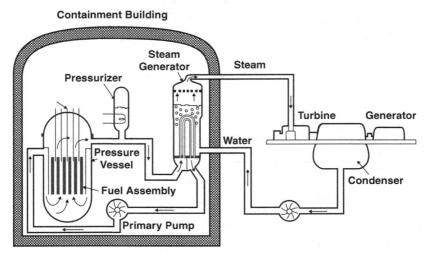

Fig. 5.   Schematic of a PWR.

Inside the steam generator, the water in the primary circuit transfers its heat to boil the water in the closed *secondary circuit*. The steam generated is dried and sent to the turbine before being condensed back to liquid water and recycled. This condensation is achieved in a large *condenser* cooled by water coming from the sea or a large river. Sometimes, the circuit cooling the condenser is a closed circuit, itself cooled by evaporation of water inside a *cooling tower*. There are also a number of auxiliary circuits, notably for the emergency core cooling system (ECCS).

The power level of the reactor is controlled by adjusting the rate at which fissions occur within the core. This is achieved by inserting or removing *control rods* in and out of the core; these rods are made of a material (e.g. boron, cadmium, or hafnium) that strongly absorbs neutrons without undergoing fission. In addition, boric acid can be added to the primary water circuit when the fuel is "fresh".

The PWR fuel is made of ceramic cylindrical *pellets* made of an oxide of enriched uranium (see later) or a mixture of uranium and plutonium oxides. These pellets are inserted inside long thin hollow tubes made of an alloy of zirconium, a metal that does not absorb neutrons and is very resistant to corrosion by hot water. Those tubes, the *cladding*, are then welded closed to constitute the *fuel pins*.

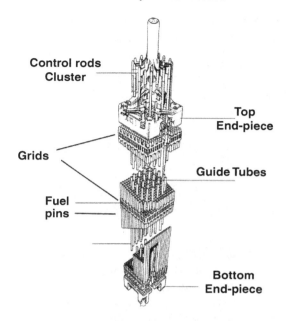

**Control rods
Cluster**

**Top
End-piece**

Grids

Guide Tubes

**Fuel
pins**

**Bottom
End-piece**

17 x 17 PWR Fuel Assembly

Fig. 6.   Typical PWR fuel assembly (*Courtesy*: AREVA).

These pins are assembled around a "skeleton" of guide-tubes to make fuel assemblies with top and bottom end-pieces. A cluster of control rods can be inserted in the guide tubes, through the action of mechanisms located on top of the pressure vessel head (Fig. 6).

4 Reactor Types

A nuclear reactor needs a mixture of fissile and fertile materials to make the fuel, a fluid to act as coolant and, most of the time, a **moderator** to slow down the neutrons. We have already mentioned the possible choice of fissile and fertile materials. For the coolant, one can choose among liquids (ordinary water, heavy water,[a] molten sodium or other metal, organic fluid, etc.) and gases (carbon dioxide or helium). The moderator must have

---

[a]In a molecule of heavy water, the hydrogen atoms are heavy: Their nucleus consists of one proton and one neutron, instead of one single proton like the ordinary hydrogen nucleus. This heavy hydrogen is called deuterium and denoted D. Heavy water is therefore D_2O. Having already one neutron, D is not a neutron absorber. Ordinary water H_2O is sometimes called "light water" to emphasize the difference.

Table 5:   Type and Net Electrical Power of Reactors Connected to the Grid and Under Construction.[4]

Type of reactor	Operating #	1/5/2014 GWe	Under construction #	1/5/2014 GWe
PWR or VVR	274	254	60	60
BWR	81	76	4	4
GCR	15	8	—	—
PHWR (*)	48	24	5	3
LWGR (**)	15	10	—	—
FBR	2	0.6	2	1.3
HTR	—	—	1	0.2
Total	**435**	**373**	**72**	**68**

Note: (*) or CANDU (**) or RBMK.

light nuclei to efficiently slow down the neutrons by successive collisions, and must not absorb too many neutrons in the process, which limits the practical choice to water (ordinary or heavy) and graphite. However, fast neutron reactors can be designed with no moderator at all.

If, in addition, you take into account the various physical–chemical states of the fuel and its different possible shapes, the number of combinations of reactor types is huge.

Between 100 and 200, different types of nuclear reactors have been designed, built and operated in the 1950s and early 1960s, the "pioneer era". But today only a handful of nuclear plant types constitute the world's nuclear fleet, as shown in Table 5. The oldest plants have a power rating in the range 300–500 MWe[b] while the newest rate from 1100 to 1600 MWe.

Chapter 2 will give a more detailed description of the reactor types.

5 The Nuclear Fuel Cycle

One cannot run a car on crude oil. In the same way, one cannot run a nuclear plant on uranium ore. The series of industrial steps necessary to prepare the nuclear fuel and manage the spent fuel after it is unloaded from the reactor core is called the nuclear fuel cycle (Fig. 7).

---

[b]MWe is a measure of the electrical power generated, as opposed to the thermal power expressed in MWth.

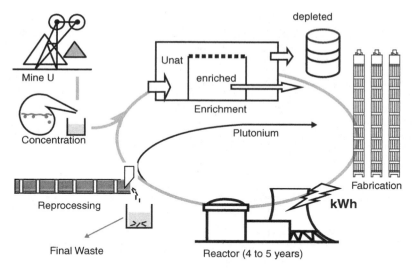

Fig. 7. The nuclearfuel cycle.

Table 6: Uranium Conventional Resources.

| Mt U | Identified | | To be discovered | |
US $/kg U	Reasonably assured	Inferred	Prognosticated	Speculative
<40	0.57	0.23		
40–80	1.95	1.00	1.70	3.74
80–130	1.01	0.65	1.11	
130–260	0.48	0.42	0.09	0.16
>260	?	?	?	3.59
Subtotal	4.01	2.30	2.90	7.50
	6.31		**10.40**	

5.1 *Uranium resources*

Uranium concentration in the continental crust of the Earth is of the order of 3 parts per million (ppm), intermediate between the common and the precious metals. Some ore bodies in Canada have concentrations above 10% in uranium, but most deposits in the world rate around 1% or less. The latest assessment of uranium resources is shown in Table 6, as compiled in the *Red Book* jointly issued in 2010 by IAEA and OECD/NEA[5]: The figures expressed in millions of metric tons refer only to conventional uranium resources. Unconventional resources, notably as by-products in phosphate deposits, exceed 22 Mt of uranium. There are over four billion tons of

Table 7:   Identified Uranium Resources as a Function of the Cost (Thousands of Tons U).

Country	<$40/kg U	<$80/kg U	<$130/kg U	<$260/kg U
Australia		1349	1662	1739
Kazakhstan	47	486	629	820
Russia	0	55	487	650
Canada	351	417	469	614
Namibia	0	7	261	518
USA	0	39	207	472
Niger	6	6	421	446
South Africa	0	186	279	372
Brazil	138	229	277	277
Ukraine	6	62	120	225
China	59	135	166	166
Uzbekistán	71	71	96	96
Total World	**681**	**3079**	**5327**	**7097**

uranium in the oceans, but so diluted as to make its industrial exploitation unrealistic in the medium term.

New assessments, published in 2012, upgrade the identified resources recoverable at a cost below or equal to $360 per kg of uranium to 7.1 million metric tons. For comparison, world uranium consumption amounted to 64,000 tons of uranium in 2011. Identified conventional resources would be able to feed the *existing* fleet of nuclear reactors for over a century, and we can expect to discover much more.

On the other hand, if a vigorous nuclear renaissance increases the world fleet from 440 to 2,000 reactors, lifetime uranium supply will become an issue, leading to the deployment of fast breeders.

Together, Australia, Kazakhstan, Russia, Canada, USA, South Africa, Namibia, Niger, Ukraine, and China account for 85% of the world's identified reserves, out of which 27% are found in Australia alone (Table 7). In 2011, Kazakhstan, Canada, Australia, Namibia, Russia, Niger, Uzbekistan, and USA accounted altogether for 97% of the 55,000 tons of uranium produced in the world.

5.2 *Exploration, mining, and concentration*

Uranium itself is not very radioactive, but some of its daughter products emit strong γ radiation. Once a potentially interesting region has been identified from geological maps, aerial surveys can detect this radiation, calling for closer local exploration to assess whether such anomalous uranium concentration qualifies as a commercially exploitable deposit.

According to its depth, uranium ore can be extracted by underground or open-air mining operations. Large and deep low-grade deposits in sparsely populated areas can also be exploited by *in situ* leaching (ISL). A number of wells are drilled into the ore body; an alkaline solution is injected into these wells to dissolve the uranium which is thereafter extracted by pumping up the solution through neighboring extraction wells.

As most deposits contain 1% uranium or less, the ore must be concentrated before transportation. In the concentration mill near the mines, ore is crushed and ground, then dissolved to form a pulp. After clarification and purification using solvents or resins, uranium is precipitated in a concentrate called *yellowcake*. Yellowcake, which contains around 75% uranium, is packed in drums and shipped to the next facility in the cycle.

5.3 *Conversion and isotopic enrichment*

Natural uranium consists mostly of two isotopes, ^{238}U (99.28%) and ^{235}U (0.72%). ^{238}U has a half-life of 4.47 billion years while ^{235}U has a half-life of 713 million years, which means that since the formation of Earth, half of the ^{238}U is still present while only 2% of the original ^{235}U still remains.

In a light water reactor (LWR), hydrogen nuclei in the water absorb too many neutrons for a chain reaction to be sustained if the uranium contains only 0.72% fissile ^{235}U. One must *enrich* the natural uranium to reach 3–5% ^{235}U before fabricating the fuel. As both isotopes have identical chemical properties, isotopic separation processes use the slight mass difference between the two isotopes.

First, the yellowcake is purified from any neutron-absorbing impurity and *converted* into uranium hexafluoride UF_6, a solid at room temperature but gaseous above 60°C. Natural fluorine has only one isotope, ^{19}F, so the UF_6 mass differences come only from the uranium isotopes. Early enrichment plants used a process called gaseous diffusion but all modern plants use the *centrifuge* process.

In a centrifuge (Fig. 8), a high narrow cylindrical bowl rotates at extremely high speed. The *feed* UF_6, noted F, is introduced at the center of the bowl. Heavier $^{238}UF_6$ molecules gather at the bottom and closer to the wall, where they are collected as *depleted* flow W, while lighter $^{235}UF_6$ molecules collect toward the top and farther away from the cylinder and are gathered as *enriched product* P. Each centrifuge increases the ^{235}U

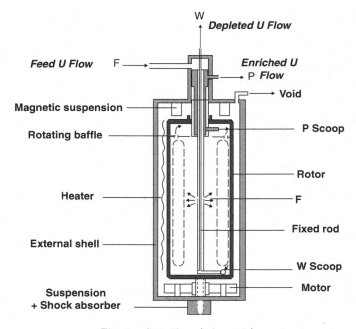

Fig. 8.   Centrifuge (schematic).

assay of the enriched flow and a *cascade of* several tens of centrifuges is needed to reach the target enrichment. As a single cascade contains very little material, enrichment plants are made of a series of many parallel mounted cascades.

The separative power of a cascade is expressed in a conventional unit called separative work unit (SWU). To produce 1 kg of 4% enriched uranium, 8 kg of natural uranium is required plus 6 SWUs. In addition, 7 kg of depleted uranium is stockpiled with an assay of 0.25% ^{235}U, which can be used later as a fuel for fast breeder reactors. To enrich the uranium needed to supply a 1 MWe PWR, one requires around 120,000 SWUs per year.

Table 8 gives the world enrichment capacities in million SWU. Figures in italics refer to gaseous diffusion, the others, to separation by centrifuge. In addition, Pakistan and Iran have small non-commercial facilities.

5.4 *Fuel manufacture (PWR)*

Preliminary operations must be first carried out: Zirconium alloy cladding tubes must be manufactured and assembly end pieces must be machined

Table 8:   Enrichment Capacities (Million SWUs/year).

Production capacity	2006	2015
AREVA France	10.8	7.5
URENCO Europe	9.0	12.0
JNFL Japan	1.1	1.5
USEC USA	11.3	3.5
URENCO USA	—	3.0
AREVA USA	—	1.0
TENEX Russia	25.0	33.0
CNNC China	1.0	1.0
Total	**58.2**	**62.5**
Demand	**48.4**	**57–63**

while UF_6 must be ***defluorinated*** and converted into UO_2 powder. Spacing grids are assembled.

UO_2 is then pressed to make cylindrical pellets, which are sintered under a reducing atmosphere to become ceramics. These pellets are machined and introduced into the cladding tubes, leaving an empty plenum at the top of the tube. The tube is pressurized with helium and a plug is welded on top, making it a completely leak-tight fuel pin. The pins are then introduced within the "skeleton" constituted by the guide-tubes held in place by the grids. Top and bottom end pieces are fitted to complete the fuel assembly, ready to be shipped to the plants after many checks and inspections.

5.5 *Open cycle or closed cycle?*

The "fresh" fuel assembly is loaded in the reactor core. It remains there for four or five years, being shuffled around the core during every reload. After this period of time, it can no longer produce power: It has lost many fissile nuclei (even though new nuclei were bred from the fertile material) and some of the fission products are neutron absorbers. In addition, its structural strength has been weakened by corrosion, creep and radiation damage. The fuel assembly is therefore unloaded and put in temporary storage in a dedicated pool of water at the nuclear plant. It is now termed ***spent fuel***.

Some countries, like Finland and Sweden, consider this spent fuel as just radioactive waste, to be disposed of as such as we shall see in Section 7

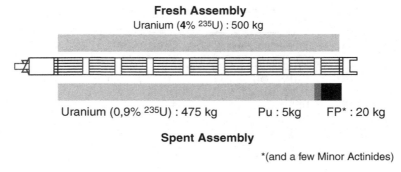

Fresh Assembly
Uranium (**4%** 235U) : 500 kg

Uranium (0,9% 235U) : 475 kg   Pu : 5kg   FP* : 20 kg

Spent Assembly

*(and a few Minor Actinides)

Fig. 9.   Content of one spent PWR fuel assembly.

of Chapter 3. They do not recycle anything, and one speaks of an **open cycle** — which means no cycling at all.

Other countries, like France and Japan, consider that spent fuel is a mixture of final waste and recyclable materials (Fig. 9): They operate a closed cycle by reprocessing the spent fuel to recover the leftover uranium and the bred plutonium and they store the properly conditioned final waste, metallic parts, fission products, and actinides, awaiting their disposal.

The recovered uranium has less than 1% 235U and must be re-enriched to make new fuel. The recovered plutonium is mixed with depleted uranium to make mixed oxides (MOX) fuel. MOX pellets are loaded in fuel assemblies identical to uranium fuel assemblies. Plutonium having higher neutron absorption, the plutonium concentration in MOX pellets is twice the 235U concentration in low enriched uranium (LEU) pellets.

5.6 *Reprocessing and vitrification*

Reprocessing operations are aimed both at recovering recyclable fissile materials and at minimizing the long term radio-toxicity of the final waste.

After one or two years in the spent fuel pool of the nuclear plant, spent fuel assemblies are shipped to the reprocessing plant in massive casks, where they are stored a few more years under water to allow for further cooling. The decrease in radioactivity, very significant in the first years after unloading from the reactor, simplifies the reprocessing operations.

At the request of the utility that owns the fuel, spent assemblies are removed from the pool to be reprocessed. **Shearing machines** are used to cutoff the end pieces and then to cut up the fuel pins into segments, about 35 mm in length, which drop into a **dissolver**. In the dissolver, hot concentrated nitric acid completely dissolves the pellets, leaving only hollow segments of metallic cladding called **"hulls"**.

End pieces and hulls are rinsed, dried, and compacted under high pressure to make metallic pancakes, which will be piled up in cylindrical stainless steel containers.

After clarification, the solution that contains nitrates of uranium, plutonium, fission products, and minor actinides is sent to the ***chemical separation*** facility. There, using solvent extraction processes, uranium and plutonium are separated from the waste products. In the ***PUREX*** process, the only industrial process, a second stage of extraction separates uranium and plutonium.

Plutonium is converted into oxide powder and temporarily stored before being shipped to the MOX fabrication plant. Uranium is either stored or sent to the conversion plant for re-enrichment.

The solution containing the waste products, fission products, and minor actinides is sent to the ***vitrification*** facility. There, the nitrates are heated in a rotating oven to become oxide ***calcines***. Those calcines are mixed with glass frit in a furnace, where they are melted and refined to produce a homogeneous liquid glass. This liquid glass is poured into a cylindrical stainless steel container, where it solidifies into a ***massive glass block*** incorporating about 14% fission products.

Advanced reprocessing processes called partitioning and transmutation (P&T) are under study to further separate long-lived minor actinides from the fission products. Separated neptunium and americium can be fissioned, but only in future Generation IV fast breeders. A modification of PUREX can prevent the complete separation of pure plutonium (COEX).

6 Economics

Nuclear plants are expensive to build and inexpensive to operate. Capital costs account for more than half of the levelized kWh cost which, therefore, is heavily dependent upon the discount rate used in the cost calculations.

The International Energy Agency (IEA) and the Nuclear Energy Agency (NEA), both part of OECD,[5] publish synthesis of future electricity cost studies carried out by their member states on a regular basis. The latest issue refers to plants being commissioned in 2015 and generating baseload electricity. A carbon tax of $30 per ton of CO_2 is included in the costs of electricity from fossil fueled plants, and decommissioning of nuclear plants is assumed to cost 15% of their construction cost. Table 9 summarizes this

Table 9:   Projected Future Costs of Baseload Electricity
($/MWh).

Region	Nuclear	Coal	Gas
DR = 5%			
USA	50 (48–51)	73 (68–15)	81 (77–91)
Pacific*	33 (29–50)	62 (54–88)	86 (67–105)
Europe	62 (50–80)	81 (68–120)	90 (85–119)
DR = 10%			
USA	75 (73–77)	90 (88–94)	88 (83–95)
Pacific*	49 (42–76)	75 (71–107)	89 (75–120)
Europe	108 (82–136)	100 (87–152)	97 (88–123)

Note: *Pacific = Japan, South Korea, and Australia.

Table 10:   Breakdown of Projected Costs.

	Investment (%)	O&M (%)	Fuel (%)	Carbon(%)
DR = 5%				
Nuclear*	60	24	16	—
Coal	28	9	28	35
Gas	12	6	70	12
DR = 10%				
Nuclear*	75	15	9	—
Coal	42	8	23	27
Gas	16	5	67	11

Note: *Nuclear costs include provisions for waste disposal and dismantling. Nuclear fuel costs include uranium, conversion, enrichment, and
fabrication.

study for real discount rates of 5% and 10%. For each region surveyed, three
figures are given in $/MWh: Median, low, and high.

With a low discount rate of 5%, nuclear power is very competitive, but
less so with 10%. Table 10 gives the breakdown of the median costs.

Of course, these calculations have made assumptions about the future
prices of coal, gas,[c] and uranium. But Table 10 explains why nuclear costs
are much less sensitive to uranium prices than fossil costs are to coal prices,
not to mention gas!

---

[c]At the time of publication, the study did not take into account the impact of shale gas
on gas prices in the North American continent.

7 Non-Proliferation

The term "proliferation" refers to the rise in the number of states in possession of nuclear weapons. The term "non-proliferation", on the other hand, refers to the political or technical means implemented to combat proliferation.

7.1 *Brief history*

Year	Country	Proliferation	Non-proliferation
1945	USA	First A bomb	
1949	USSR	A bomb	
1952	United Kingdom USA	A bomb First H bomb	
1953	USSR USA	H bomb	Atoms for peace
1956	UN		Creation of the IAEA
1957	United Kingdom	H bomb	
1960	France	A bomb	
1963	USA/USSR/UK		Moscow Treaty (to limit tests)
1964	China	A bomb	
1967	China	H bomb	
1968	France World	H bomb	Non-Proliferation Treaty (NPT)
1974	India IAEA Exporters	A "Test"	"Trigger" list London Suppliers Group NSG
1990	Iraq	Clandestine program	Gulf War
1991	South Africa		Dismantled weapons + joined NPT
1995	Former Soviet Union World		Weapons returned to Russia NPT extended indefinitely
1997	IAEA		Additional protocol
1998	India	H bomb	
1999	Pakistan	A bomb	
2003	Pakistan, Libya, North Korea, ...	A.Q. Khan "Bazaar"	
2006	North Korea Iran	A bomb Enrichment crisis	Multilateral negotiations...

NB Israel is credited with possessing nuclear weapons but has not performed any nuclear test.

The United States first tried to protect its military nuclear monopoly by preventing any transfer of nuclear knowledge. When proliferation occurred in the USSR anyway, President Eisenhower changed tack and allowed other countries access to reactor technology in exchange for their commitment to using the technology for peaceful applications only.

During the Cold War, the United Kingdom, then France, and at last China joined the Nuclear Weapon States (NWS). In 1968, the Non-Proliferation Treaty (NPT), attempted to freeze the situation by recognizing five legitimate nuclear powers but no more. In exchange, the nuclear powers undertook to reduce their arsenal and give free rein to civil technology transfers. IAEA, setup by the UN in Vienna in 1956, was entrusted with the task of overseeing the peaceful use of nuclear materials. Many countries joined the NPT.

In 1974, India, which did not sign the NPT, broke the growing consensus by carrying out a "peaceful explosion", using plutonium produced in a heavy water reactor supplied by Canada. Exporting nations then agreed to regulate "sensitive" exports. In 1991, a similar shock was felt with the discovery of the extensive clandestine nuclear program of Iraq, a country that had signed the NPT. As a result, the powers and inspection capabilities of the IAEA were reinforced.

With the end of the Cold War in 1990 and the disintegration of the USSR, the Russian Federation became the sole inheritor of the former nation's military nuclear power status. Significant disarmament programs were implemented in the NWS (except for China). With the indefinite extension of the NPT in 1995, optimism prevailed.

But in 1999, Pakistan, India's rival since partition in 1948, crossed the "nuclear Rubicon". Then, in 2003, Libya revealed the existence of a black market of nuclear weapons technology managed by A. Q. Khan, the "father of the Pakistani bomb". This, in turn, led to the discovery of a clandestine enrichment program carried out by Iran in violation of its international commitments as a party to the NPT. The most recent event was the test of a nuclear device by North Korea in 2006.

Despite the Iranian and North Korean crises, the NPT, backed up by IAEA inspections, now forms the universally acknowledged basis for all nuclear commerce. The only areas of resistance are Israel, India, and Pakistan and now, Iran and North Korea.

7.2 *Proliferation and civilian nuclear technologies*

So far, no country has ever proliferated by diverting nuclear materials or facilities under IAEA safeguards. Nuclear fission cannot be uninvented: Any country ready to devote enough financial and technical efforts, and willing to pay the political price, can make weapons. North Korea is not the most advanced country in terms of industry or technology. Nevertheless, it has been able to successfully explode a nuclear bomb. Proliferation is much more a matter of political will than a matter of technology, but, to make weapons, highly concentrated fissile materials, highly enriched uranium (above 90% ^{235}U), or "weapons grade" plutonium are needed.

Not all areas of the nuclear industry are "sensitive" in terms of proliferation. Isotopic enrichment using centrifuges has become the most common "proliferating" technology: it is easy to hide a clandestine facility and not too difficult to divert an existing facility from its purely civilian purpose. On the other hand, a reactor which can be refueled online has the capability to produce nearly pure ^{239}Pu (which requires a very short irradiation time). Used in combination with spent fuel reprocessing, such a reactor would offer the most efficient way to build up an arsenal.

It is very unlikely that a would-be proliferator could make bombs using plutonium recovered from spent LWR fuel. As noted by Pellaud,[6] nobody has ever done so. Conversely, surplus weapon-grade plutonium from the Cold War inventories can be "demilitarized" by burning it as MOX fuel in civilian power plants.

8 Conclusion

Modern wind turbines bear little relationship to antique windmills, and modern supercritical coal plants with elaborate scrubbers bear little resemblance to their ancestors. Only photovoltaics and nuclear fission are really new energy technologies. Both are based on the "new" physics discovered in the early years of the 20th century. Nuclear fission is a very complex way to boil water, but nuclear power is today, with run-of-the-mill hydraulics, the only way to generate baseload electricity with minimum emission of greenhouse gases.

References

1. W. Marshall (1986). *Nuclear Power Technology*. Clarendon Press.
2. ENDFB Files. CSEWG Document ENDF-102 Report BNL-90365-2009 Rev.1, July 2010.
3. R. M. E. Diamant (1982). *Atomic Energy*. Ann Harbor Science.
4. OECD (2010). *Uranium 2009: Resources, Production and Demand* (OECD).
5. OECD IEA/NEA (2010). *Projected Costs of Generating Electricity* (OECD IEA/NEA).
6. B. Pellaud (2002). *J. Nucl. Mater. Management*, 30(3).

Chapter 2

Current and Future Fission Power Plants

Bertrand Barré

Institut National des Sciences et Techniques Nucléaires
CEA Saclay, 91191 Gif sur Yvette Cedex, France
bcbarre@wanadoo.fr

In 2010, after two decades of quasi-stagnation, nuclear power was back as a main source to generate reliable baseload electricity with very low greenhouse gas emissions. The March 2011 accident at the Fukushima Daiichi nuclear power plant has delayed this rebound by a couple of years. For this "renaissance" to be sustainable, there are prerequisites: No severe accident, an accepted way of disposing of long-lived waste, etc. If nuclear power grows rapidly, new types of "Generation IV" systems will have to be deployed to better use the fissile material resources.

1 Introduction

In 2014, 435 nuclear reactors operating in 30 countries supplied around 5% of the world's primary energy and 12% of the world's electricity. Table 1 lists, by country, the reactors in operation and under construction on January 1st, 2016, by number of units and capacity in GWe, as well as the amount of electricity generated during the year 2014, in billion kilowatthours (TWh) and in percentage of total electricity production in each country. France, has by far the highest percentage of electricity (77%) and the second highest total amount of electricity (418.0 TWh) produced by nuclear power.

Although nuclear fission was discovered in 1938 and the first nuclear "pile" went critical in December 1942, nuclear power production started only in the mid 1950s. The number and size of nuclear reactors grew rapidly in the 70s and 80s, but the world nuclear capacity grew very slowly

Table 1:   Nuclear Reactors in Operation and Under Construction, June 30, 2015.

Country	Nuclear electricity generation 2014 (billion kWh)	Reactors operable January 1, 2016		Reactors under construction January 1, 2016	
		% e	No.	MWe net	No.	MWe gross
Argentina	5.3	4.0	3	1627	1	27
Armenia	2.3	30.7	1	376	0	0
Belarus	0	0	0	0	2	2388
Belgium	32.1	47.5	7	5943	0	0
Brazil	14.5	2.9	2	1901	1	1405
Bulgaria	15.0	31.8	2	1926	0	0
Canada	98.6	16.8	19	13553	0	0
China	123.8	2.4	30	26849	24	26885
Czech Rep	28.6	35.8	6	3904	0	0
Finland	22.6	34.6	4	2741	1	1700
France	418.0	76.9	58	63130	1	1750
Germany	91.8	15.8	8	10728	0	0
Hungary	14.8	53.6	4	1889	0	0
India	33.2	3.5	21	5302	6	4300
Iran	3.7	1.5	1	915	0	0
Japan	0	0	43	40480	3	3036
S, Korea	149.2	30.4	24	21677	4	5600
Mexico	9.3	5.6	2	1600	0	0
Netherlands	3.9	4.0	1	485	0	0
Pakistan	4.6	4.3	3	725	2	680
Romania	10.8	18.5	2	1310	0	0
Russia	169.1	18.6	35	26053	8	7104
Slovakia	14.4	56.8	4	1816	2	942
Slovenia	6.1	37.2	1	696	0	0
S, Africa	14.8	6.2	2	1830	0	0
Spain	54.9	20.4	7	7002	0	0
Sweden	62.3	41.5	9	8849	0	0
Switzerland	26.5	37.9	5	3333	0	0
Ukraine	83.1	49.4	15	13107	0	0
UAE	0	0	0	0	4	5600
UK	57.9	17.2	15	8883	0	0
USA	798.6	19.5	99	98990	5	6218
China Taipei	*40.8*	*18.9*	*6*	*4927*	*2*	*2700*
WORLD	**2.411 billion kWh**	**c 11.5 % e**	**439 No.**	**382.547 MWe**	**66 No.**	**70,335 MWe**

from 1985 to 2005, and mostly in Asia, as shown in Fig. 1, from the PRIS database of IAEA.[1]

   From 2005 to 2010, nuclear power was experiencing what the media called a "renaissance". Even though the percentage of the world's electricity

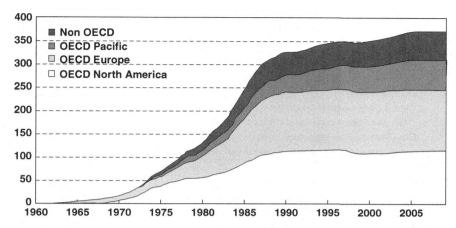

Fig. 1. World nuclear generating capacity (GWe) 1960–2009.
Source: IAEA 2010.

generated by nuclear power decreased from 18% in 2000 to 14% in 2009, and the total amount of electricity generated by nuclear power decreased from 2,626 TWh in 2005 to 2,558 TWh in 2009, this is only the result of two decades of low construction rate and the progressive shutdown of the oldest plants. The severe accident which occurred in Fukushima in March 2011 led to the progressive shutdown of all the Japanese nuclear power plants (NPPs) and the return of Germany to its previous nuclear phase-out policy. However, the global consequence of the Fukushima accident is likely to be a delay of a couple of years in the "renaissance" of nuclear power. More and more countries are now planning to introduce nuclear into their energy mix and the number of reactors under construction, was 27 in 2005, is currently 72 (in May 2014).

There are at least three reasons for this renaissance. From an economic point of view, nuclear power is immune from the erratic variations of oil and gas prices, and almost insensitive to the price of uranium, which accounts for less than 5% of the total cost of energy from a nuclear plant. From a geopolitical point of view, uranium resources are well distributed across the continents in countries with very different political regimes. Last but not least, from an environmental point of view, nuclear power emits almost no greenhouse gases (GHGs) into the atmosphere over its complete lifecycle (Fig. 2).

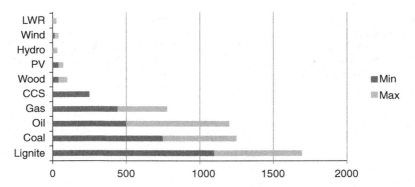

Fig. 2.   Lifecycle GHG emissions, in grams of CO_2 equivalent for 1 kWh of electricity.

Note: Ranges reflect differences in assessment technology, conversion efficiency, assessment boundary, etc.

Source: D. Weisser, IAEA 2006.[2]

Table 2:   Summary of Reactor Types.

Type	Fuel	Coolant	Moderator	Cycle
PWR	LEU or MOX	Ordinary water	Ordinary water	Indirect
BWR	LEU or MOX	Ordinary water	Ordinary water	Direct
GCR	Natural U or LEU	Carbon dioxide	Carbon dioxide	Indirect
Candu	Natural U	Heavy water	Heavy water	Indirect
RBMK	LEU	Ordinary water	Graphite	Direct
FBR	$(U, Pu)O_2$	Liquid sodium	—	Indirect
HTR	LEU or $^{233}U/Th$	Helium	Graphite	Direct

2 "Generations" of Nuclear Reactors (Fig. 3)

One usually describes the development of nuclear plants in terms of generations. Generation I plants were the early reactors of the pioneer era, a series of prototypes of increasing sizes and very varied designs, with no standardization. Most of them are now shut down and decommissioned. Generation II consists of the reactors presently in operation, mostly light water reactors (LWRs). Generation III reactors are presently under construction: Modernized versions of LWRs, they embody significant improvements in safety and protection to prevent any severe accident. While Generation I plants are being decommissioned, Generation II plants are being operated and Generation III plants are under construction. International efforts are currently under way to develop Generation IV nuclear systems, which are designed to match the requirements anticipated to be in force around 2040–2050.

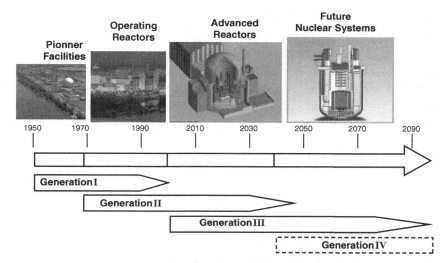

Fig. 3. Successive Generations of nuclear reactors.

3 Plants in Operation (Generation II Reactors)

A summary of different types of nuclear reactors is shown in Table 2 and discussed in more detail in the following sections. See also Fig. 3.

3.1 *Pressurized water reactors (PWRs)*

PWRs (or VVRs[a] for those designed in the Soviet Union or in Russia today) currently dominate the world market. This type was described in Chapter 1. In VVRs, the fuel pins are assembled with a triangular pitch instead of a square one, and the steam generators are horizontal rather than vertical. Otherwise, the technology of the PWRs and VVRs is very similar.

3.2 *Boiling water reactors (BWRs)*

Like the PWR, the BWR is cooled and moderated by ordinary water, but it uses a direct cycle: the water boils at the outlet of the core and the dried steam goes directly to the turbine while the liquid part is recycled within the pressure vessel. Instead of being assembled in a cluster, the control rods are cross-shaped and inserted in the core from below. The fuel assembly is similar to that of PWRs, but with an external wrapper which separates fuel channels inside the core. Together, PWR and BWR are called lLWRs.

---

[a]From the Russian: *Vodo-Vodianoï Energuetitcheski Reaktor.*

3.3 *Gas-cooled reactors (Magnox, AGR, HTR)*

Gas-cooled reactors played a significant role at the beginning of nuclear power, notably in the UK and in France, but they survive only in the former. They use graphite as a moderator and carbon dioxide (CO_2) as coolant. The Magnox are fueled with rods of metallic natural uranium while the more recent advanced gas-cooled reactors (AGRs) are fueled with low enriched uranium dioxide.

A more advanced version of gas-cooled reactors, the high temperature reactor (HTR or sometimes the High Temperature Gas Reactor, HTGR) has an all-refractory core cooled by very hot helium gas (750°C or more). Several prototypes have been operated, in two families called "prismatic" or "pebble bed" according to the shape of their fuel elements, but commercial maturity is expected only in Generation IV reactors.

3.4 *Heavy water reactors (PHWR or Candu)*

The pressurized heavy water reactor (PHWR) is moderated by a huge tank of warm heavy water,[b] which surrounds an array of horizontal pressure tubes inside which the fuel assemblies are cooled by hot pressurized heavy water. This "primary" heavy water goes to a collector and then to a steam generator very similar to that of a PWR, where ordinary water boils, the resulting steam being sent to the turbine.

As heavy water absorbs very few neutrons, the fuel pellets are made of natural uranium oxide. Therefore their fuel does not need enrichment. But heavy water is expensive to manufacture and the core is much larger than an LWR's, which means higher capital investment. Canadian deuterium uranium (Candu) reactors are mostly used in Canada and India.

3.5 *Light water graphite reactors*

In a High Power Channel-type Reactor (RBMK is the Russian language acronym), arrays of vertical pressure tubes are installed throughout a huge graphite massif. Inside each tube, a fuel assembly is cooled by boiling ordinary water. The pressure tubes are connected to collectors where steam is separated and carried to the turbine. RBMKs were only located within the Soviet Union proper (Russia, Ukraine and Lithuania) because their design

---

[b]In heavy water, the hydrogen (proton) is replaced by a deuteron (proton plus neutron) and is sometimes written D_2O.

enables them to produce weapon-grade plutonium. Unit 4 of Chernobyl, an RBMK, was the origin of the worst ever nuclear accident in April 1986.

3.6 *Fast breeder reactors (FBRs)*

In a FBR, neutrons are not slowed down by a moderator. The fissile materials inventory in their core must be larger than in a thermal reactor, but they can breed i.e. turn more fertile nuclei into fissile nuclei than they burn fissile nuclei in the process. By this means, through multiple recycling, they can make use of all the potential energy of the uranium resource. Reactors today use less than 1% of this energy. FBRs are fueled with a mixture of uranium and plutonium oxides and are cooled by molten sodium. Several prototypes have been (or are) in operation, but the technology is demanding and commercial maturity is expected only for Generation IV, when the threat of uranium scarcity makes them attractive.

4 Generation III[3]

In Chapter 3, we shall describe briefly the accident which occurred at Chernobyl in 1986. The main lesson drawn from this accident was: "Nevermore!". This led to the design of a new generation of nuclear power plants, still in the LWR family, but with one additional constraint: even the most severe accident — full core meltdown — should not result in large releases of radioactivity to the environment.

To meet this requirement, new systems were developed to prevent accidents, but also to mitigate their consequences outside the plant itself. Generation III plants can be both PWR and BWR, but there are two basic options:

- "Evolutionary" plants voluntarily limit the number of design innovations, in order to benefit the most from the large amount of experience gained from presently operating reactors (of Generation II). They use both active and passive safeguard systems.
- "Revolutionary" plants are more innovative, with special emphasis on passive systems, but they lack the benefit of extensive experience.

Active systems require operator action and usually need some external power source (electricity, compressed air, etc.), while passive systems use "spontaneous" phenomena like natural convection, gravity, and evaporation/condensation.

AREVA's 1650 MWe EPR is a good example of an evolutionary plant, while Toshiba–Westinghouse's 1,200 MWe AP 1000 is a revolutionary design. All Generation III plants have also been reinforced against aircraft crashes following the extensive damage produced by airplanes on September 11, 2001 in New York.

4.1 *AP 1000*

The AP 1000 is an extrapolation in size (1,200 MWe) of the AP 600 reactor, designed by Westinghouse during the 80s to be a revolutionary reactor, relying much more on passive safeguard systems than on active ones.

The primary circuit (Fig. 4) comprises two loops each of them being equipped with one large steam generator, one hot leg and two cold legs, and two primary pumps, wet canned, directly connected to the bottom

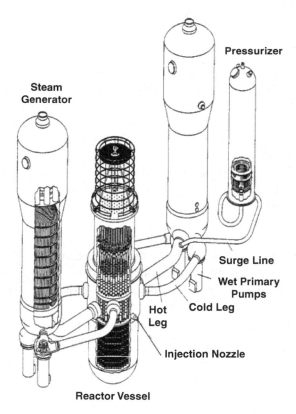

Fig. 4.   AP 1000 primary circuit.

Fig. 5. Artist's view of the Westinghouse AP1000.

end of the stream generators. The containment is made of a single steel shell located within a concrete chimney. This chimney is topped by a large annular water tank which gives the plant its characteristic profile (Fig. 5). Within the containment is a large in-containment refueling water storage tank (IRWST), two core makeup tanks and two accumulators (there is no emergency injection pump).

If a breach develops in a primary pipe causing a loss of coolant accident (LOCA), the water from all these tanks would flood the bottom of the containment shell automatically, creating a kind of "pool", and the water level would reach the bottom part of the steam generators. As a result, the breach would be under water. The core would therefore be cooled, using natural convection, via the breach itself and the decay heat would be transferred to the pool.

The water in the pool would then evaporate, but the steam would condense on the wall of the steel shell and the condensed water would flow back into the pool, thus compensating for the evaporation. During the first hours after the accident, the shell would be externally cooled by water flowing by gravity from the tank atop the chimney. The time needed to empty the tank is enough for the decay heat of the core to decrease to the point where it can be evacuated by the air flowing between the shell and the chimney by natural circulation. The whole sequence would happen passively with a minimum of intervention from the operators, and without the need for power beyond batteries.

In case of core meltdown, the corium[c] would be solidified back *in situ* by flooding the vessel pit.

The designers of AP 1000 have especially endeavoured to limit on-site construction time: Some 250 large modules are prefabricated in the factory and set into the shell's "open top" by powerful cranes. The shell itself is raised, section after section during the assembly. The vendor hopes to limit on-site construction time to 36 months from pouring the first concrete to loading the fuel.

4.2 *The evolutionary power reactor (EPR)*

In their joint 1993 directives, the French and German Safety Authorities had made clear the importance they placed on using the "return of experience" as a safety assurance mechanism. They said in essence: "if your new model is so innovative that the cumulative experience of both the French and German operating fleets is no longer fully relevant, be prepared for some serious discussions during the safety analysis procedure". This message was well understood by the designers: ***EPR is purposely evolutionary*** and its only significant innovations are dedicated to enhance its safety to the Generation III level.

With this proviso, EPR is an accomplished marriage between the most recent PWR models from both side of the Rhine River, viz. N4 and Convoy. Similarly, it is a 4-loop PWR, but its rating is somewhat higher (1,650 MWe) than its predecessors.

It has four emergency core cooling trains, each with a 100% capability to bring the reactor to safe shutdown. These four trains are located in four separate safeguard buildings, so as to combine functional redundancy with geographical separation, thus avoiding common mode failures.

The auxiliary buildings are built close to the reactor building and share with it the same extra-thick basemat, a very seismic-resistant setup (Fig. 6).

EPR has a double containment. An inner thick hull made of prestressed concrete with an internal steel liner, and an outer hull made of heavily reinforced concrete. A similar hull also protects the fuel building as well as

---

[c]The corium is the mixture of molten metals and oxides resulting from the meltdown of the core inside the pressure vessel. It is similar to lava.

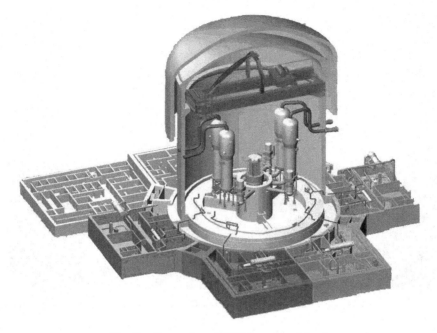

Fig. 6.   Open view of EPR's nuclear island.

two of the four safeguard buildings (the other two being located on each side of the reactor building). Such a disposition guarantees outstanding protection against any external aggression, be it natural or malevolent.

Six diesel groups housed in two separate buildings located on either side of the reactor buildings supply the emergency electric power supply when required. Each massively constructed and waterproof diesel building houses two "normal"diesel groups, each with a 100% capability, and another group of a different technology dubbed "SBO" which stands for station black-out. The SBO diesel would be called upon if a common mode failure were to affect simultaneously all four normal diesel groups. Fuel tanks for the diesel groups are also located within the buildings.

In case of a severe accident with full core meltdown and corium escaping from the reactor pressure vessel, the following sequence is anticipated:

• The corium spreads into the refractory spreading zone where it is cooled down, first by the water flowing passively from the large IRWST (1,800 m^3) around it, then by the aspersion system located at the containment ceiling.

Fig. 7. Panoramic view of the Taishan site in China where two EPR are being built.
Source: AREVA.

- This aspersion system is powered by the emergency diesel groups or by the SBO diesels, and is cooled by a dedicated system in addition to the safety-grade cooling system
- Pressurized core meltdown is prevented by opening dedicated valves on top of the pressurizer (2×900 t/h).

The risk of a hydrogen explosion is prevented through some 60 passive catalitic recombiners located within the inner containment. This inner hull can resist a pressure of around 10 bars. Residual leakages through the inner containment penetrations would gather in the inter-containments annular space and be filtered before release to the stack.

Figure 7 shows the site in China where two EPR are being built.

4.3 *Other Gen III plants*

The PWR branch comprises, in addition to the EPR and the AP 1000, the VVR AES 92 of Rosatom, the APWR of Mitsubishi and the ATMEA jointly developed by AREVA and Mitsubishi. Whether the Korean APR1400 is Generation III is debatable because its proposed management of a molten core is not fully convincing.

The BWR branch is shorter: The ABWR of General Electric and Hitachi, GE's ESBWR and AREVA's Kerena.

As of May 2014, APWR, ATMEA, ESBWR and Kerena have not yet been ordered. Two AES 92 units are ready to commence operating in India. The other units are still under construction: two ABWR in Taiwan, two APR1400 in South Korea, four AP 1000 in China and four in the USA, and four EPR in Finland, France, and China, respectively.

5 Generation IV Nuclear Systems[4]

As explained above, Generation IV nuclear systems are only at the design stage today. This new generation will probably not be deployed until the world fleet of operating nuclear plants has reached a size where serious questions arise about the ability to provide sufficient uranium as fuel over the anticipated lifetime of the plants. This date is impossible to forecast with any degree of certainty.

Obviously, any breakthrough, any introduction of a brand new reactor species must be prepared decades in advance. With this kind of delay, one can hope to design not only "reactors", but whole nuclear systems better suited to meet the needs of the 2040s or 2050s, which may not be identical to the criteria which led to the present selection and deployment of reactor types.

Philosophically, this is a real change. Instead of designing reactors optimized for today's market — like Generation III, one must try to anticipate what the market will be like four decades from now, to imagine the best possible concepts to answer these distant demands and to launch today all the actions necessary, including R&D, to make this desired future possible. In fact, one tries to replace the past "natural" spontaneous selection by a formalized *process*. This process is being carried out currently through two main international initiatives.

5.1 *The Generation IV International Forum (GIF)*[5]

At the initiative of the US Department of Energy, since 1999, 14 countries and the European Union have worked together to select a few model concepts for future nuclear systems, and to define and perform the R&D necessary to make some of them ready for possible commercialization after 2040. Criteria for *formal* selection include **Sustainability** (fissile resources utilization, waste minimization, proliferation resistance, and physical protection), **Safety & Reliability** (radio-protection, reactivity control, heat

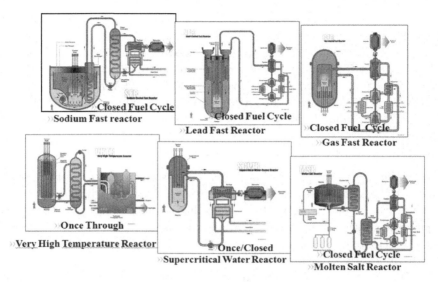

Fig. 8.   The six concepts selected by GIF.

removal, mitigation features) and **Economics**. A first phase was open to the unbridled creativity which led to revisiting all former reactor designs and inventing a few more. The result was a list of roughly 120 concepts. Then, within the Generation IV International Forum, each concept was passed through the sieve of a very careful review. Six of them survived the ordeal.

The six model concepts finally selected were (Fig. 8):

- Sodium-Cooled Fast Reactor (SFR) System;
- Lead Alloy-Cooled Fast Reactor (LFR) System;
- Gas-Cooled Fast Reactor (GFR) System;
- Very High Temperature Reactor (VHTR) System;
- Supercritical Water-Cooled Reactor (SCWR) System;
- Molten Salt Reactor (MSR) System.

All these systems do not have the same timeframe for development. One could reinvent Superphénix — or, rather its successor design EFR — in a comparatively short time. The VHTR and SCWR could be developed at the prototype level rather quickly. The GFR is further away in time, and the MSR even further down the road.

The importance of the "sustainability" criterion is clear from the fact that at least four and possibly five of these systems are based on closed fuel cycles to better use the fissile resources (naturally occurring uranium or even available depleted uranium). In fact, three of the systems are fast breeder reactors.

The VHTR and (later) the GFR are aimed at not only the generation of electric power but also the cogeneration of power and hydrogen with the aim of making synthetic transportation fuels using fission power. This is certainly necessary if we want nuclear power to contribute significantly to the reduction of greenhouse gas emissions on a world scale.

Finally the future of nuclear power may not be limited to these six species. Some intriguing concepts are being developed like the AHTR, a molten salt-cooled HTR. Some teams continue to work on the Accelerator Driven subcritical systems, which could be used for the transmutation of minor actinides, if this was deemed useful at some point.

5.2 *International Project on Innovative Nuclear reactors & Fuel Cycles (INPRO)*[6]

In 2000, the IAEA initiated the INPRO Project in which 15 (now over 20) member states have worked to define "User Requirements" for innovative nuclear energy systems in the area of Economics, Sustainability and Environment, Safety, Waste Management, Proliferation Resistance, and some cross-cutting issues. The time horizon of this exercise is 2050. They have also developed a methodology of assessment for such systems. Phase 1A of INPRO was completed in June 2003, and the methodology is being tested on six "national cases". Started entirely as an extra-budgetary initiative, INPRO is now partly funded by the IAEA's regular budget.

Although they are based on similar analyses and motivations, the work of GIF and INPRO are not identical: the GIF partners are mostly suppliers, and their work will steer the R&D, whereas INPRO expresses mostly the requirements of potential future users. Each group is quite aware of the other's results, and they have worked in cooperation on some issues (like proliferation resistance).

5.3 *Fast breeders (SFR, LFR, GFR)*

Fast breeders have fascinated reactor designers since the beginning of the 1950s because they are able, through successive recycling, to use almost all the potential energy of the uranium resource, a 100 times more than the LWRs. Even better, they can use the depleted uranium, leftover from the enrichment process, as a source of fuel.

Because liquid water is too effective a moderator, the choice of a coolant for an FBR can only be between liquid metals and gases. The only significant experience that can be utilized comes from sodium-cooled SBRs. Used between 300°C and 500°C, sodium is an excellent coolant, with good compatibility with stainless steels if the sodium is pure. However, sodium burns in air and reacts strongly with water.

In order to prevent any reaction between the activated sodium of the primary circuit and the water going to the turbines, current SBRs have an intermediate circuit of inactive sodium. This additional circuit and the use of "noble" non-reactive materials in contact with the sodium increases the investment costs compared to LWRs. SBRs will therefore be developed only when a shortage of uranium becomes a key factor in the economics of nuclear reactors.

Industrial size SBR plants have been put into operation in France (Superphénix, 1,200 MWe) and the Federation of Russia (BN 600), as well as significant demonstration plants in France, Great Britain, Japan, Kazakhstan, and the United States. Smaller experimental facilities are in operation in India and China.

To power some of their nuclear submarines, the Russians have developed LFR fast reactors cooled by liquid lead or a lead/bismuth (Pb/Bi) alloy. Lead does not have the chemical reactivity of sodium but it must be kept above 400°C to remain liquid. Lead is very dense and as opaque as sodium and can corrode stainless steel under some circumstances. The Pb/Bi eutectic alloy is liquid at much lower temperature than pure lead but bismuth is not abundant and under neutron irradiation it produces very radio-toxic polonium.

Gas cooled GFR have attractive characteristics, being both breeders and high temperature reactors, but removing the heat from radioactive decay after loss of gas pressure would be a serious problem and the core of such reactors still remains to be designed.

If one chooses eventually to partition and transmute some long-lived nuclides (see Chapter 3) in order to manage the high level radioactive waste, FBRs would be well suited to the task, with their high neutron flux and the possibility of locally adjusting the neutron spectrum to a specific capture resonance.

Because they are much more efficient in the use of fissile resources, FBRs occupy the lion's share of the Generation IV concepts.

5.4 *Other Gen IV Concepts*

The VHTR is only an upgrade of the older HTR design. The HTR was developed around a very innovative fuel of British origin, the "Coated Particle". This finely divided fuel, the proportion of which can easily be adjusted within a bulk graphitized moderator, allows for very refractory cores, cooled by high temperature helium. This, in turn, allows for high thermal efficiency, comparable to the best fossil fueled plants. By mixing different particles, one can adjust to almost any fuel cycle.

The HTR equivalent of a fuel assembly exists in two varieties: a prismatic block with coolant channels or a spherical nodule or "pebble".

Their large thermal inertia and significant margins *vis-à-vis* their operating conditions (as part of the space propulsion program NERVA[7], several HTR reactors have heated their hydrogen coolant above 2,500°C) make HTRs very safe reactors. On the other hand, their low power density results in high specific investment costs.

After promising demonstrations, HTRs had a false start in the beginning of the 1970s in the USA and during the 1980s in Germany. They are presently being reconsidered as small modular reactors with direct helium cycle cooling. Their small size allows for passive decay-heat cooling after shutdown. Small experimental HTR units are operating in Japan and China and aim to demonstrate the direct production of hydrogen from nuclear heat, by catalytic water splitting.

The supercritical water reactor (SCWR) concept is a PWR in which the primary pressure and temperature keep ordinary water in the supercritical state, neither liquid nor gaseous. Given the high nuclear safety standards, a pressure of 24 MPa or more would be very demanding on the materials of the primary circuit.

The last concept is the MSR, which can be designed with either a thermal neutron spectrum (with molten fluorides and graphite moderator) or a fast neutron spectrum (with molten chlorides). The fuel constitutes its own coolant.

A very small MSR facility operated in Oak Ridge from 1965 to 1969, demonstrating the specific qualities of this concept: Fuel cycle flexibility, online reprocessing, and recycling (which avoids transportation) and with a small breeding potential. With such a small amount of operating experience, any commercial deployment is far away. MSR designs often use the thorium cycle (see Section 7).

6 Small and Medium (or Modular) Reactors (SMR)

The history of nuclear power plants has been characterized by a quasi-constant escalation of unit ratings: 70, 120, 300, 440, 600, 900, 1,100, 1,300, 1,450 and now 1,650 MWe.

Why is there such a trend? The reason is because the cost of generating nuclear power is very sensitive to the size effect: the larger the plant, the more it costs to build but the less the cost of a given unit of electricity output (kWh). This size effect, obvious from 70 to around 1,000 MWe, may be less clear above this rating because the more powerful plants, built more recently, must comply with more stringent regulations. Meeting these new standards requires additional investments which may balance the size effect. In France, for instance, moving from the large series of thirty four 900-MWe plants to the series of twenty 1,300-MWe units came with the "un-twinning"[d] of units which significantly increased the construction cost.

Economics of scale is not the only driver of this escalation: it is almost as lengthy and difficult to get local acceptance and all the administrative clearances for a 200-MWe plant as for a 1,600-MWe unit. For the same final output, it is simpler, cheaper, and less time consuming to do this once instead of eight times.

On the other hand, for a grid to remain stable, the power of each node must not exceed one-tenth of the total grid capacity. To connect 1,600 MWe nuclear units, one needs a very large national grid or well-interconnected neighboring grids. This escalation of the unit ratings makes it more difficult to introduce nuclear power in a "newcomer" country.

In addition, instead of building a big plant on a site, with all the climatic hazards, construction site constraints, and difficulties in ensuring quality control in the open air, it would be more comfortable to **fabricate** big modules in a plant and have only to **assemble** the final reactor on-site. Or indeed, one could even build a complete small reactor in a controlled environment and simply ship it to the final use site.

That is why, periodically, plant designers or their potential customers reconsider the question: Should we not revisit the concept of "Medium" reactors, with power ratings from 300 MWe to 700 MWe, or even of "Small" reactors with a power rating of less than 300 MWe? Since the mid-1980s, there has also been the hope of balancing the size effect through

---

[d]900 MWe units are twinned. Each pair share a number of auxiliary buildings while 1,300 MWe units stand alone.

"Modularity". If, on the same site, one builds small identical nuclear units, progressively assembled like so many Lego blocks, the first units will start generating power and, therefore, supplying cash, when one starts to finance the remaining units. It is even possible, though far from certain, that modularity might accelerate the licensing procedures.

Every 10 years or so, there is a peak of interest in SMRs, but — until now — this has not resulted in any such plants being ordered and built. We are today witnessing such a peak in interest.

In 2012, the IAEA published a comprehensive review[8] of all the SMR designs. Those in operation are usually outdated designs but some of those under development are quite innovative. A symbol of the present peak of interest for SMRs is the recent decision by the US Department of Energy to spend US $450 million to subsidize two reactor vendors, Babcock & Wilcox (B&W) and NuScale, a subsidiary of the Fluor Corporation linked with Rolls-Royce, to help them finance two demonstration projects and their licensing cost viz. B&W's "mPower": reactor which would have a power rating of 180 MWe, while NuScale is developing a very small, modular, 45MWe unit. Both are PWRs. As a result, Westinghouse has placed its own SMR project on the backburner "for lack of customers"[9] (Fig. 9).

The Russians have several SMR projects under construction. The South Koreans are developing SMART, a SMR dedicated to the cogeneration of electricity and freshwater. In Argentina, CAREM looks like a naval propulsion reactor. The Chinese have started building a high temperature SMR, and in France there is a project termed FLEXBLUE for a nuclear plant 100 m under water with a rating 160 MWe. And the list goes on...

So, there is no lack of designs available but the real question is: is there a market for SMRs?

There seems to be an internal market in the Russian Federation: in Siberia, there are several mining or industrial communities located hundreds of miles away from any electrical grid. As a matter of fact, the Russians have already four old small nuclear plants supplying electricity and heat to the town of Bilibino.[10]

There might also be a case for SMRs — paradoxically — in the USA. There, power is produced and distributed by a very large number of small "utilities", often not well interconnected. There are also a lot of medium sized coal plants which must be shut down in the near future because they pollute too much: a few years ago, this seemed to constitute a good opportunity for SMRs, but given the present low price of shale gas in the USA, they cannot compete. One should also remember that in the 1980s, US

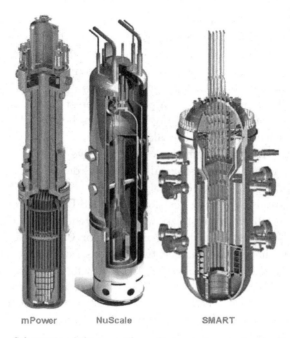

mPower   NuScale   SMART

Fig. 9.   Schematic of three small, medium reactors under development.

utilities demanded in the Utilities Requirements Document (URD) that nuclear plants have a rating around 500 MWe. Such a demand was the origin of Westinghouse's AP 600 and General Electric's SBWR. However both those medium sized plants never found any customers and the AP 1000, four units of which are now under construction in the United States, is an uprating to 1,200 MWe of the late AP 600.

In contrast, there are the newcomer countries, some of which have too small a grid to accommodate large nuclear plants. Those countries might be interested in medium sized plants if there were already some positive experience from demonstration SMRs in the vendor country.

A last "niche" could be considered: supplying power to islands like Hawaii or La Réunion. But this is only a niche market. So, the jury is still out in terms of judging the commercial prospects of SMRs.

7 The Thorium Cycle

Almost all reactors today operate on the uranium cycle, burning fissile ^{235}U complemented by plutonium "bred" from ^{238}U (see Chapter 1). But if ^{235}U is the only fissile nucleus available in nature, ^{238}U is not the only natural fertile isotope. One can also breed fissile ^{233}U by capturing neutrons in ^{232}Th. One neutron capture is followed by two successive β^- decays: the parallelism between the two cycles is striking.

$$^{238}\text{U} + {}^{1}\text{n} \rightarrow {}^{239}\text{U} \rightarrow {}^{239}\text{Np} \rightarrow {}^{239}\text{Pu},$$
$$\beta^- \qquad \beta^-$$
$$^{232}\text{Th} + {}^{1}n \rightarrow {}^{233}\text{Th} \rightarrow {}^{233}\text{Pa} \rightarrow {}^{233}\text{U}.$$

Thorium, with atomic number 90, is two elements below uranium, with atomic number 92, in the periodic classification of elements. It is a slightly radioactive metal belonging to the actinide family. It was discovered by Berzelius in 1828 and named after Thor, son of Odin and god of thunder in the Nordic mythologies. Its atomic mass is 232.12. Thorium dioxide ThO_2, is used to make very refractory ceramics.

The use of thorium in reactors was considered very early in the game. The high value of η, the number of emitted neutrons after one absorption by uranium 233, makes breeding possible even with thermal neutrons, while breeding with the uranium/plutonium cycle is only possible in fast neutron reactors. But in any thorium-fueled reactor it is necessary to "jump-start" the first cores with a fissile isotope (^{235}U or ^{239}Pu), before being able to recycle ^{233}U and enter the *real* thorium cycle.

Almost all the reactor types have been tested with thorium fuel as long as highly enriched uranium (HEU) was commercially available. But when the HEU trade was banned for the sake of non-proliferation, the thorium cycle was abandoned everywhere but in India.

As a matter of fact, between 1974, (the date of the "peaceful" Indian nuclear explosion) and 2009, India was cut off from the international nuclear trade, including uranium imports. Condemned to isolation and with limited domestic uranium resources but large thorium reserves, India bet on thorium for its long term nuclear future.

When thorium is mentioned nowadays, it is usually in association with MSRs. To best use its qualities, a MSR must be coupled to a chemical plant which purifies its liquid fuel online. This "reprocessing" is especially efficient with the thorium cycle because it allows you to separate protactinium 233 and let it decay outside the reactor core without capturing neutrons to produce useless ^{234}U. A small MSR demonstrator was tested in the 1960s and the concept is one of the 6 Generation IV models. In the past, all MSR designs were thermal neutron reactors, but a French CNRS team is promoting a fast neutron MSR which is attractive on paper.[11] Recently, China is considering building a new thorium-fueled MSR demonstrator.

Thorium is likely to complement uranium in the future, but its promoters tend to exaggerate its advantages.

In the Earth's crust, thorium is 3 to 4 times more abundant than uranium, and there are large resources in Brazil, India, China, USA and Madagascar. On the other hand, there is practically no thorium in seawater.

You need a large number of successive neutron captures to go from thorium to americium or curium. This reduces the long term potential radio-toxicity of the spent fuel as compared to the uranium cycle. This, however, is not enough to remove the need for a deep geological repository.

One hears sometimes that the thorium cycle would be less "proliferation-prone" than the uranium cycle. This argument is not convincing. While ^{233}U is a better fissile material than ^{235}U, its "contamination" with ^{232}U, the decay products of which emit high energy gamma rays, would be much more a nuisance for the fabrication of civilian fuel assemblies (which require tons of material) than for the fabrication of bombs (which require only kilos) which can be done before the build-up of ^{228}Th, the "father" of the radioactive decay products.

Despite numerous past studies and the realization of a few demonstration projects, the thorium cycle completely lacks the 6 decades of R&D which went into the development, industrialization and optimization of the uranium cycle. The future will tell if and how well this handicap can be overcome.

Glossary

AES 92	Russian Generation III PWR
AGR	Advanced Gas-cooled Reactor (British Generation II)
AP 1000	Toshiba-Westinghouse Generation III PWR
AP 600	Predecessor to AP 1000
APR 1400	Korean Modern PWR
APWR	Japanese Generation III PWR
ATMEA	French–Japanese Generation III PWR
BWR	Boiling Water Reactor
CAREM	Argentine Small PWR
CNRS	French Research Organization
EPR	French Generation III PWR
ESBWR	US Generation III BWR

FBR	Frat neutron Breeder Reactor
FLEXBLUE	French small PWR project
GCR	Gas-cooled Reactors
GFR	Gas-cooled FBR
GHG	Greenhouse effect Gas
GIF	Generation IV International Forum
HTR, HTGR	High Temperature Gas-cooled Reactor
IAEA	International Atomic Energy Agency
INPRO	International Project on innovative Nuclear systems
IRWST	In-reactor Water Storage Tank
Kerena	French generation III BWR
LFR	Lead (or lead-bismuth)-cooled FBR
LOCA	Loss of Coolant Accident
LWR	Light Water Reactor (PWR or BWR)
mPower	Small US PWR project
MSR	Molten Salt Reactor
NPP	Nuclear Power Plant
NuScale	Small US PWR project
PHWR, Candu	Canadian Deuterium Uranium (Pressurized Heavy Water Reactor)
PWR	Pressurized Water Reactor
R&D	Research & Development
RBMK	Light Water-cooled Graphite-moderated Reactor
SBO	Station Black Out
SBWR	Predecessor to ESBWR
SCWR	Supercritical Water Reactor
SFR	Sodium-cooled FBR
SMART	Korean small PWR for power and desalination
SMR	Small and Medium (or Modular) Reactors
URD	US Utilities Requirements Document
VHTR	Very High Temperature Reactor
VVR	*Vodo-Vodianoï Energuetitcheski Reaktor* Russian name for PWR

References

1. World Nuclear Association, from IAEA PRIS. Accessed on January 13, 2016 at https://www. iaea.org/pris.
2. D.Weisser, PESS/IAEA (2006.). A guide to life-cycle greenhouse gas (GHG) emissions from electric supply technologies. Accessed at https://www.iaea.

org/OurWork/ST/NE/Pess/assets/GHG_manuscript_pre-print_versionDan-ielWeisser.pdf

3. B. Barré (2012). *Third Generation Nuclear Plants.* C.R. Physique 13. Académie des Sciences, Elsevier, May 2012.

4. B. Barré *et al.* (2016). Nuclear Reactors Systems, A Technical, Historical and Dynamic Perspective. EDP Sciences.

5. A Technology Roadmap for Generation IV Nuclear Energy Systems. GIF 002-00 December 2002

6. INPRO in Brief. IAEA, August 2012

7. JW Simpson. Nuclear Power from Underseas to Outer Space. ANS 1995

8. Status of Small and Medium Sized Reactor Designs. A Supplement to the Advanced Reactors Information System. IAEA, September 2012.

9. Interview of Dany Roderick to the Pittsburgh Gazette, February 2014

10. Power Reactor Information System. Accessed on January 2015 at www.iaea.org/PRIS/.

11. Th. Auger *et al.* The CNRS Research Program on the Thorium Cycle and the Molten Salt reactors. PACEN/CNRS June 2008.

Chapter 3

Nuclear Safety and Waste Management

Bertrand Barré

Institut National des Sciences et Techniques Nucléaires
CEA Saclay, 91191 Gif sur Yvette Cedex, France
bcbarre@wanadoo.fr

The two issues of public concern about nuclear power are the consequences of possible severe accidents and the guarantee of very long term protection against the radiations emitted by nuclear waste. Both for safety and waste management, containment is the name of the game. Safety is a process of continuous progress where lessons drawn from past accidents are used to reduce more and more the probability of severe accidents, and to mitigate their consequences on human health and the environment. Low level radioactive waste is safely disposed of in licensed repositories in many countries. The disposal of civilian high level, or long-lived medium level waste is not yet fully implemented, but the scientific community agrees that the best solution is the disposal in deep geological repository. Finland and Sweden are rather close to building such repositories. The Oklo phenomenon gives a strong indication that such a solution can be safe much longer than the time for the radioactivity to become negligible.

1 Nuclear Safety

1.1 *Introduction*

Nuclear **Security** encompasses Safety, Radiation protection, Malevolence prevention, and post-accidental intervention.

- Nuclear **Safety** is the set of technical dispositions and organizational measures concerning the design, construction, operation, shutdown, and decommissioning of nuclear facilities, as well as the transportation of radioactive substances, in order to prevent accidents and mitigate their consequences.

- **Radiation Protection** is the set of rules, procedures and means of prevention and control which seek to prevent or reduce the harmful effects of ionizing radiation on workers and the general public.

Safety is a prime responsibility of the plant operator. In any country which is party to the International Safety Convention, a national Safety Authority, fully independent of the plant operator, authorizes the operation, controls the compliance with safety regulations, and has the power to force the shutdown of facilities it deems no longer safe to operate.

1.2 *Barriers and defense-in-depth*

The nuclear industry extracts, produces, handles, treats and transports huge quantities of radioactive substances. The radioactive nuclei in those substances emit radiation, which can be harmful to living tissue if absorbed at high enough doses. The whole basis of nuclear safety is therefore to **contain** the radioactive substances behind suitable *"barriers"*, able to stop the radiation so that it remains harmless. Containment is the name of the game everywhere: In reactors, fuel cycle facilities, transport casks and waste storage facilities or repositories. In the following, let us focus on reactors, taking the pressurized water reactors (PWRs)[a] as an example.

The radioactive nuclei resulting from fission within the fuel pellets are contained inside the leak-tight metallic cladding of the fuel pins, which constitutes the first barrier. If some pins start to leak, the radioactive elements will still be contained within the steel envelope of the closed primary circuit: a second barrier. In addition, the primary circuit is fully contained within the thick reinforced concrete walls of the reactor building, usually made of double enclosures. This containment building is the third (and ultimate) barrier preventing the radioactivity of the core from reaching the plant operators and the public-at-large. As long as at least one of these barriers remains intact, no radioactivity will escape to the environment.

Nuclear plants embody the concept of **defense in depth** to minimize the consequences of possible human errors or material failures:

- The plant must be designed safely, built accordingly, and correctly operated and maintained by well trained specialists.

---

[a]See Glossary at the end of Chapter 2 for acronyms.

- In case of abnormal operation, redundant automatic *protection* systems will be activated to put the plant back on track.
- If these first two lines of defense fail, redundant automatic *safeguard* systems will be activated to bring the plant to safe shutdown, with at least one of the barriers still intact.
- If, and only if those three lines were to fail, will radioactivity escape: This is a *severe accident* and emergency measures must be taken to protect the population outside the plant.

To maintain the integrity of the first two barriers in a PWR, one must control the chain reaction to prevent reactivity accidents, and one must dissipate the heat produced within the fuel, *including* the residual *after heat* still generated by the radioactive decay of fission products once the chain reaction has been stopped.

Reactivity accidents are only minor contributors to the total risk of fuel meltdown in light water reactors (LWRs) because of two negative feedback coefficients, intrinsic to the core design. If the core temperature increases, ^{238}U nuclei absorb more neutrons, while the moderation by water decreases: Both phenomena create a negative *temperature coefficient* and the chain reaction stops by itself. Similarly, if the water density decreases by leakage or excessive boiling, the moderation is less efficient and this negative *void coefficient* stops the chain reaction. In contrast, at Chernobyl the strong **positive** void coefficient of the particular reactor there, (an RBMK[b]) was one of the main triggers of the accident.

On the other hand, as much as 90% of the risk of fuel meltdown comes from the possible failure to properly cool it down. This is what happened in Three Mile Island 2 (TMI-2) and Fukushima, even though automatic shut down had operated perfectly. This explains why a series of diverse safeguard systems are devoted to the function of Emergency Core Cooling.

1.3 *The International Nuclear Events Scale[1] (INES)*

Any abnormal event occurring in a nuclear facility is reported to the Safety Authority. Significant events are made public in the interest of transparency and full disclosure.

Nuclear plants are complex facilities, and explaining the degree of significance of a given event to the public and the media is difficult. To convey

---

[b]See Glossary at the end of Chapter 2.

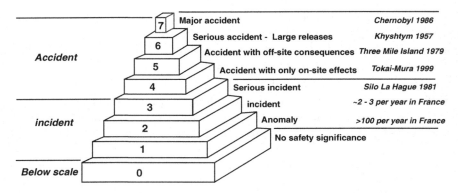

Fig. 1.   The international nuclear events scale[1] (INES).

the message, a logarithmic event scale (see Fig. 1) analogous to the well-known Richter scale, which describes the severity of earthquakes, has been defined internationally, ranging from 1 to 7. Below level 1, events must still be reported but they have no safety significance. Levels 1–3 qualify as *incidents* of increasing significance. *Accidents* start at level 4 and reach 7 in case of a *major* accident. Chernobyl and Fukushima are to-date the only major nuclear accidents. TMI-2 was rated at level 5. In France, with 58 operating NPPs, more than 100, level 1, and two or three level 2 incidents are declared every year.

Fortunately, very few reactor *accidents* have been recorded since the beginning of the use of nuclear power:

- Windscale (UK), 1957, level 5
- SLX (USA), 1961, level 5
- Lucens (Switzerland), 1969, level 4
- St Laurent 1 (France), 1969, level 4
- Three Mile Island 2, 1979, level 5
- St Laurent 2 (France), 1980, level 4
- Chernobyl 4 (Soviet Union), 1986, level 7
- Fukushima 1,2 & 3 (Japan), 2011, level 7

The most severe accident in a non-reactor facility occurred in 1957 at Khystym in the Urals where the explosion of a fission products container contaminated a vast area. This was rated at level 6. The criticality accident at Tokai–Mura (rated level 4) was also a non-reactor event.

The early accidents had very little public impact, contrasting with the three we shall now describe.

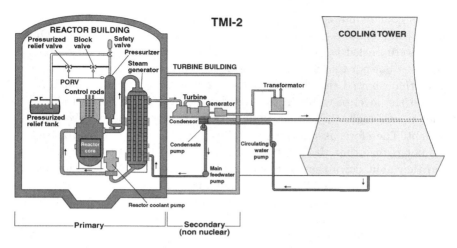

Fig. 2.   Schematic view of the TMI 2 nuclear plant.

1.4 *Three Mile Island, March 28, 1979*[2]

The Three Mile Island nuclear plant is located on a sand island in the middle of the Susquehanna river, 15 miles from Harrisburg, the capital of Pennsylvania. The plant comprises 2 PWR units rating 800 MWe and 900 MWe, built by Babcock & Wilcox. TMI 1 started operation in 1974 and TMI 2 at the end of 1978.

The accident began about 4:00 a.m. on March 28, 1979, when the plant, running at 97% power, experienced a failure in the secondary, non-nuclear section of the plant. The main feedwater pumps supplying water to the steam generators (Fig. 2) stopped running, which prevented the steam generators from removing heat. First the turbine, then the reactor automatically shut down. Immediately, the pressure in the primary system began to increase. In order to prevent that pressure from becoming excessive, the protection system of the plant opened the relief valve located at the top of the pressurizer (PORV). The valve should have closed when the pressure decreased, but it did not. As a result, cooling water poured out of the stuck-open valve to the relief tank. Responding to the loss of cooling water, high-pressure injection pumps automatically pushed replacement water into the reactor system. As water and steam escaped through the relief valve, cooling water surged into the pressurizer, raising the water level in it.

As coolant flowed from the core through the pressurizer, the instruments available to reactor operators provided confusing information. Operators were misled by a signal indicating that the valve had been ordered closed — but not that the valve *was actually* closed. There was no instrument that showed the level of coolant in the core. Instead, the operators judged the level of water in the core by the level in the pressurizer, and since it was high, they assumed that the core was properly covered with coolant. Therefore, as alarms rang and warning lights flashed, the operators did not realize that the plant was experiencing a loss-of-coolant accident, and they shut down the emergency water injection. Steam then formed in the reactor primary cooling system. Pumping a mixture of steam and water caused the reactor cooling pumps to vibrate, and the operators shut them down as well.

Without forced circulation, water and steam were stratified and the top of the core was cooled only by steam. Because adequate cooling was not available, the nuclear fuel overheated to the point at which the zirconium cladding (the long metal tubes which hold the nuclear fuel pellets) ruptured and the fuel pellets began to melt. It was later found that about one-half of the core melted during the early stages of the accident. In addition, when the reactor's core was uncovered, a high-temperature chemical reaction between water and the fuel cladding had produced hydrogen gas.

About two and a half hours after the start of the accident, radioactive water overflowed from the relief tank and gathered at the sump in the bottom of the reactor building. This water was automatically pumped to the auxiliary building, where its radioactivity triggered alarms. This is when the operators started to understand what was really happening. At 6:22 a.m. operators closed a block valve between the relief valve and the pressurizer. This action stopped the loss of coolant water through the relief valve. However, superheated steam and gases blocked the flow of water through the core cooling system and they had to fight until the evening to be able to restore water circulation in the primary circuit.

Although the TMI-2 plant suffered a severe core meltdown, the most dangerous kind of nuclear power accident, it did not produce the worst-case consequences that reactor experts had long feared. In a worst-case accident, the melting of nuclear fuel would lead to a breach of the walls of the containment building and release massive quantities of radiation to the environment. But this did not occur as a result of the Three Mile Island accident.

Many studies of the radiological consequences of the accident have been conducted by several agencies and independent organizations. Estimates are that the average dose to about 2 million people in the area was only about 0.01 mSv.[c] Compared to the natural radioactive background dose of about 1–1.25 mSv per year for the area, the collective dose to the community from the accident was negligible. The maximum dose to a person at the site boundary would have been less than 1 mSv.

Today, the TMI-2 reactor is permanently shut down and defueled, with the reactor coolant system drained, the radioactive water decontaminated and evaporated, radioactive waste shipped off-site to an appropriate disposal site, reactor fuel and core debris shipped off-site to a Department of Energy facility, and the remainder of the site being monitored. Completing the clean-up took 14 years.

The main lesson learned from the accident was the importance of the "human factor": It is not enough that the plant itself is safe; the instrumentation of the plant must be clear and the operator must understand the instrumentation in detail. Many of the improvements resulting from TMI 2 are aimed at improving this communication, viz.

- Removing ambiguities from the instrumentation.
- Selecting alarms in the control room to show only the most urgent problems.
- Forbidding any operator action during the first 20 min of an accident.
- Periodically training operators on simulators.

But many other improvements were also implemented in the wake of TMI such as:

- Making some critical components more reliable.
- Establishing the Institute of Nuclear Plants Operators (INPO) in the USA.
- Installing hydrogen recombiners within the reactor building.
- Putting high efficiency filters in the venting stack, etc.

It was concluded that implementing the "lessons learned" from TMI reduced the probability of a future severe accident by two orders of magnitude.

---

[c]A sievert (SV) is the International System of Units (SI), unit of radiation dose. 1 Sv represents the equivalent biological effect of the deposit of a joule of radiation energy in 1 kg of tissue.

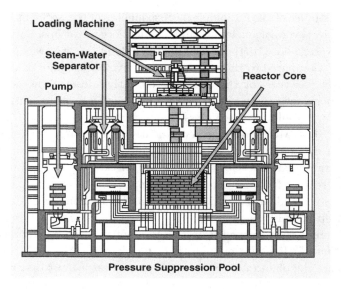

Fig. 3.   Schematic view of a RBMK.

1.5 *Chernobyl, April 28, 1986*[3]

The Chernobyl nuclear plant was located 80 km north of Kiev, the capital of Ukraine which was then a part of the Soviet Union. The plant contained four "RBMK" units rated at 925 MWe each, and two additional units were under construction. Unit 4 had been operating since March 1984.

The RBMK (Fig. 3) consists of a huge cylindrical pile of graphite crossed by vertical metal tubes. Within each tube, one fuel assembly is cooled by boiling ordinary water. The core is so large that control rods need 22 sec to be fully inserted.

In a western type LWR, whether PWR or BWR, water is used both to cool down the core and to slow down the neutrons to keep the chain reaction running. If some water is lost through leakage or over-boiling, the chain reaction stops by itself (although you still must remove the decay heat). In an RBMK, the graphite slows the neutrons down: If some water is lost, neutrons are still slowed down, but they are less absorbed by the hydrogen nuclei, and the chain reaction accelerates: This is called a positive void coefficient. At reduced power, the RBMK is not stable.

On April 28, 1986, just before the planned periodic shutdown of unit 4, its operators were preparing to carry out a safety experiment, for which they needed to incapacitate some safeguard systems. For this experiment, the reactor should have been operating at a stable reduced power during the preceding 18 hours. In order to supply power to the grid, the operators were not able to meet this condition, but they decided to start the experiment nevertheless.

Because of the **Xenon Effect**,[d] the reactor started to shut itself down and, to keep the required power level, operators removed all the control rods from the core, an operation that is strictly forbidden. Those control rods had graphite extensions. After complete withdrawal from the core, it is the graphite at the end of the control rods which re-enters the core first, and accelerates the reaction instead of quenching it.

The experiment went amok. Because it was not cooled enough, the water started boiling excessively, and because of the positive void coefficient, power escalated exponentially in the core. The control rods scram only accelerated the phenomenon. The heat was released so quickly and violently that all the water turned instantly into steam. This steam explosion overturned the 2,000 ton slab covering the core, tearing away the metal tubes like pieces of velcro tape, and blowing open the reactor roof. Part of the core was projected on the roof of neighboring building, starting about thirty fires. Overheated and now in the open air, the graphite pile took fire and the issuing thermal column sent all the volatile radioactive products into the stratosphere.

More than 10 days of heroic efforts were necessary to quench the fire. Following the jet stream, the radioactive aerosols (the so-called "cloud") flew over Europe contaminating the soil when and where some rain precipitated the radioactive elements.

The Chernobyl accident, the worst in the history of nuclear power, was rated at level 7 on the INES scale.

The health consequences of the accident have been scrutinized by the scientific community and many sociologists. The 2005 report of Forum Chernobyl, under the aegis of WHO, UNDP, UNEP and IAEA was the first exhaustive picture of the situation. UNSCEAR's 2011 report essentially confirmed the Forum's findings, viz.

- About 50 deaths occurred from acute radiation syndrome among the emergency teams responding to the accident;

---

[d]Radioactive Xenon 135, a huge absorber of neutrons, is the decay product of Iodine 135, a direct fission product. Xenon production is therefore not synchronized with the variations of the plant power level. After a power reduction, the amount of Xenon in the core increases, before decreasing by radioactive decay. If the Xenon "peak" is too high, its absorbs too many neutrons and shuts the reactor down.

- 20 deaths resulted among the approximately 7,000 patients (mostly children) who developed thyroid cancers in Belarus, Ukraine and Russia;
- 5 million people live in areas that were "contaminated", and among them 100,000 people receive annual doses higher than the level authorized for the public (1 mSv per year) but still lower than the natural dose in many areas of the globe;
- 350,000 persons were evacuated (out of whom 240,000 were moved without any benefit for their health);
- a 30 km radius zone was created where access is still forbidden today.

Independent of the radiation, the evacuation proved to be very traumatizing and resulted in mental disorders, despair, "paralyzing fatalism", etc.

To try to assess the long term effects, one is compelled to use statistical calculations which are known to be pessimistic. Among the 600,000 most irradiated individuals, inhabitants of Pripyat and early "liquidators", no more than 4,000 are expected to die prematurely from radiation-induced cancers. But it is impossible to distinguish them with any certainty from among the 150,000 deaths expected from "natural" cancers.

The accident itself was too specific to RBMKs for technical lessons to be applicable to other reactor types, but a more general lesson was learned: It is not acceptable under any circumstance to contaminate vast expanses of land and to have to evacuate people for a significant amount of time. This caused a change in the very philosophy of reactor design.

Before Chernobyl, designers defined a so-called Design Basis Accident (DBA), as a serious accident that the plant should survive, but not the most severe accident possible. The belief was that accidents more severe than the DBA had such a low probability of occurrence that such risks should be tolerated. In case of such a "beyond DBA" accident, measures would have to be taken outside the plant site.

For the nuclear plants designed after 1990, called Generation III plants (see Chapter 2), the rule is that the DBA must be the most severe accident possible. Even in case of a full core meltdown, the radioactivity escaping from the plant should be low enough so as not to require lengthy evacuation of people, nor significant condemnation of the agricultural use of the land.

1.6 *Fukushima Daiichi*[4]

The origin of the TMI accident was an ambiguous instrumentation signal which led to a severe operator error. The origin of the Chernobyl accident was an experiment carried out on a reactor which was unstable at low power, combined with several violations by the operators. The origin of the Fukushima accident was a very violent seismic event (magnitude 9 on the Richter scale) followed by a tsunami far exceeding the wave height against which the nuclear plant was protected.

On March 11, 2011, the Tohuku earthquake, the strongest ever recorded in Japanese history, did not significantly damage the 14 nuclear reactors located in the affected area, but it destroyed the electric grid in many locations including near the Fukushima Daiichi nuclear plant. The 11 nuclear reactors then in operation were automatically and successfully shut down, and in sites like Fukushima Daiichi, where the high voltage lines were destroyed, the emergency diesel generators started, as expected, to supply the power necessary to cool down the reactor cores and remove the heat from decaying radio-isotopes. Forty minutes later, a huge tsunami destroyed a large strip of the north-east coast of Honshu Island, around the town of Sendai, destroying roads and houses and erasing whole villages as far as 5 miles inland.

In Fukushima Daiichi, the tsunami caused by the earthquake flooded and totally destroyed the emergency diesel generators, the seawater cooling pumps, the electric wiring system and the DC power supply for Units 1, 2, and 4, resulting in loss of all power — except for an external supply to Unit 6 from an air-cooled emergency diesel generator. In short, Units 1, 2, and 4 lost all power; Unit 3 lost all AC power, and later lost DC power before dawn on March 13, 2011. Unit 5 lost all AC power.

The tsunami did not damage only the power supplies. The tsunami also destroyed or washed away vehicles, heavy machinery, oil tanks, and gravel. It destroyed buildings, equipment installations and other machinery. Seawater from the tsunami inundated the entire building area and even reached the extremely high pressure operating sections of Units 3 and 4, and a supplemental operation common facility (Common Pool Building). After the water retreated, debris from the flooding was scattered all over the plant site.

Unable to properly remove the heat from decaying isotopes, the cores overheated: the vapor pressure rose within the pressure vessels and the water levels dropped to the point where the top part of the fuel assemblies was no longer covered with liquid water. The fuel cladding reached the temperature where zirconium reacts with the water vapor, releasing hydrogen. The oxidized cladding released into the vapor, all gaseous and volatile radioactive species it had previously contained. The contaminated vapor reached the pressure suppression pool (wetwell) where condensation

B. Barré

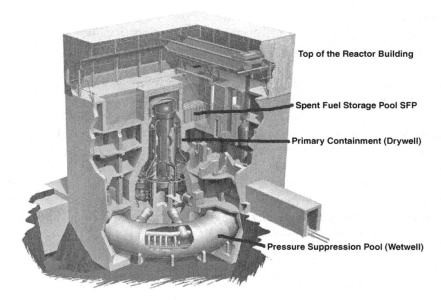

Top of the Reactor Building

Spent Fuel Storage Pool SFP

Primary Containment (Drywell)

Pressure Suppression Pool (Wetwell)

Fig. 4. Schema of a BWR mark 1.

lowered the pressure for a time, and then reached the containment vessel itself (drywell). To save the integrity of the containment vessel, the operators released the vapor to the stack, together with the radioactive gases and particles as well as the hydrogen. But the hydrogen found its way to the top part of the reactor building, where the spent fuel storage pool SFP is located. There, the hydrogen exploded in contact with the air, destroying the top of the buildings (Fig. 4).

The fuel for Unit 4 had already been unloaded previously and transferred to the SFP. However, the top of Unit 4 was also blown away by an hydrogen explosion. It was found two months later that this hydrogen came from Unit 3 as both units shared the same stack.

All the fuel of Unit 1 had melted down, passed through the bottom of the pressure vessel, RPV, and gathered on the concrete floor of the primary containment. The situation in Unit 3 was similar although to a lesser extent. The fuel of Unit 2, partly molten, is probably still within the RPV, but the containment building is leaking.

Months of heroic effort were necessary to regain control of the site following the accident, and only in December 2011 had the situation stabilized, with the cooling water now kept below 50°C. A significant amount of very radioactive water from Unit 2 had leaked into the Pacific Ocean where it was quickly diluted. From March 12 to March 15, 2011 the radioactive plumes were also blown toward the ocean, but then the wind changed direction and a 50-km long and 10-km wide band of land spreading north-west from the plant site became seriously contaminated.

A lot of work has already been done on the site, as of May 2014 including: mapping of the whole area for radioactive contamination, clean-up of the debris, connection to the grid, reinforcement of Unit 4 SFP, placing a new cover building on Unit 1, building a new fuel removal facility over Unit 4, transferring the spent fuel from Unit 4 SFP to another facility, etc. Decontamination has also been carried out outside the site on roads, schools, houses, fields and parts of the forest close to villages. But the full clean-up of Fukushima might require up to 40 years.

The most nagging problem is the accumulation of contaminated water stored on-site. About 300 m^3 of underground water is seeping daily into the lower parts of the plant buildings, where it gets contaminated. It mixes with the 400 m^3 of water which is injected daily to keep cooling the cores. Part of that water leaks to the ocean but most of it goes through a cesium removal facility and is either injected back in the plants or stored in huge tanks. The quantity of water stored in these tanks amounts to 400,000 m^3.

Very early, before any radioactive release, the Japanese authorities had evacuated the population within a 20 km radius from the damaged plant. As a result, no detectable health damage from radiation appears to have affected the population.[5] As underlined by UNSCEAR, this applies only to health effects from radiation: "the most important health effect is on mental and social well-being, related to the enormous impact of the earthquake, tsunami and nuclear accident, the forced or voluntary evacuation, and the fear and stigma related to the perceived risk of exposure to ionizing radiation. Effects such as depression and post-traumatic stress symptoms have already been reported". In June 2011, a health survey of the local population (the Fukushima Health Management Survey) was initiated. The survey, which began in October 2011 and is planned to continue for 30 years, covers all 2.05 million people living in Fukushima Prefecture at the time of the earthquake and reactor accident.

Twelve workers were estimated to have received absorbed doses to the thyroid from iodine-131 intake alone in the range of 2–12 gray.[e] They incur

---

[e]The **gray** (Gy) is a unit of the absorbed **radiation** dose in the International System of Units (SI). It is defined as the absorption of one joule of energy by one kilogram of matter.

an increased risk of developing thyroid cancer and other thyroid disorders. More than 160 additional workers received doses currently estimated to be over 100 mSv, predominantly from external exposures. To quote UNSCEAR again, *"Among this group, an increased risk of cancer would be expected in the future. However, any increased incidence of cancer in this group is expected to be indiscernible because of the difficulty of confirming such a small incidence against the normal statistical fluctuations in cancer incidence".*

In view of the amount of radioactivity released, the Fukushima accident is classified as level 7 on the INES scale. However the amount of radiation released at Fukushima is at least 10 times smaller than the radioactivity released during the Chernobyl accident, and the contaminated area is far smaller.

1.7 Lessons learned from Fukushima: The "Stress Tests"

Outside Japan, the Fukushima accident had a number of consequences on nuclear programs. Germany decided almost overnight to shut down eight nuclear plants and go back to the total phase-out decided in 2001 to take place in 2022. A few other countries (Italy, Taiwan, Switzerland) decided to cancel future nuclear projects. The other countries operating nuclear plants kept operating them and a handful of newcomers have decided after March 2011 to adopt nuclear power (Turkey, Poland, Saudi Arabia, Belarus, Bangladesh, Jordan, Vietnam). As of May 2014, the impact of the Fukushima accident appears to be to delay the "nuclear renaissance" by a couple of years.

More important, following Fukushima, the countries operating nuclear facilities have launched additional safety studies to assess the robustness of those facilities against aggressions of a level higher than those which were the basis of their original design. In the European Union, those studies — called stress tests — were decided by the European Council less than 15 days after the accident and have been uniform throughout the EU members. The Reports issued by the National Safety Authorities were even submitted for peer review by the Authorities of the other European countries.

As a result, comprehensive improvement programs have been started to:

1. Reinforce plant protection against external aggressions (earthquakes, flood).
2. Better secure the emergency supplies of cooling water and electric power.

3. Limit the release of radioactivity after an accident.
4. Reinforce the crisis management organization, locally and centrally.

In France, a number of safeguard systems have been identified as a "hard kernel" to be specially protected against aggression as their survival can prevent a serious accident from becoming severe. A special Action Force of experts was also constituted, with all the required mobile intervention materials, to be able to help nuclear plant operators anywhere in France in less than 24 hours.

2 Radioactive Waste Management and Decommissioning

2.1 *Waste categories*

One distinguishes several types of radioactive waste, according to two criteria:

- The activity level i.e. the intensity of radiation emitted: this governs the level of shielding required to provide adequate protection against radioactivity,
- The radioactive half-life of the products contained; this determines how long they are potentially harmful and must be contained.

These two criteria are opposite for a given isotope: If a radioactive element is highly radioactive, it decays very quickly and is therefore short-lived (and *vice versa* for a long-lived nuclide). What complicates matters is that a waste package very seldom contains a single radionuclide.

Radioactive waste is usually divided into three categories:

1. *Class A*: Low- and intermediate-level waste **LLW** with a short half-life (less than 31 years). Its radioactivity (β and γ *emission*) will be comparable to natural radioactivity in 300 years. It comes not only from nuclear power plants, but also from hospitals, laboratories, industry, etc.
2. *Class B*: Low- and intermediate-level (α) waste with a long half-life (several thousand years and more) **LL-ILW**. For example, the hulls remaining after spent fuel reprocessing.
3. *Class C*: High-level waste **HLW** giving off heat for several hundred years. This class emits α, β and γ radiation. It includes unprocessed spent fuel or glass containers from reprocessing and vitrification (see Chapter 1).

2.2 *Radioactive waste disposal*[6]

Class A waste constitutes the bulk of the waste volume but is only a small fraction of the radioactive waste in countries operating nuclear reactors. These wastes are disposed of in dedicated surface repositories, licensed in many countries.

The countries selecting an open cycle approach must dispose of their spent fuel as final waste. As spent fuel assemblies are not designed to contain radioactive products for millennia, they must be encapsulated in specially designed containers. Sweden and Finland have developed a high-integrity container made of pure copper. Loaded with several spent fuel assemblies gathered in a steel basket, and sealed with very careful friction welding, these copper waste packages will be disposed of in deep geological repositories dug in granite. In granite, "geological" water is chemically reducing and will not corrode the copper containers.

Spent fuel reprocessing produces two waste packages: Containers of compressed metallic waste of class B and containers of glass, class C. Both will be disposed of in deep geological repositories (clay, salt, tuff, and crystalline rocks are under consideration by various countries). A number of *in situ* underground laboratories are used to characterize the selected strata, usually some 500 m deep. Table 1 describes the status of HLW disposal in various countries, as of 2010.

Most countries have not yet chosen between open or delayed closed fuel cycles. They must keep their spent fuel in dry storage facilities, either centralized or at the power plant site, waiting for a decision concerning

Table 1: Options for Deep Geological Disposal.

Country	Waste type	Geology	Underground laboratory	Repository construction	Repository operation
Belgium	Both	Clay	Yes	2025	2040
Finland	Spent Fuel	Granite	Yes	2012	2020
France	HLW	Clay	Yes	2016	2025
Germany	Both	Salt			2035 ?
Netherlands	HLW				>2110
Spain	Spent fuel				2050 ?
Sweden	Spent fuel	Granite	Yes	2015	2023
Switzerland	Both	Clay	Yes	~2040	2050
UK	HLW				2040
U.S.	Spent fuel	Tuff	Yes	Suspended	?

Note: "?" reflects uncertainty. "Both" means HLW and spent fuel.

their disposal. The Netherlands, for instance, has licensed such a facility for more than 100 years. Only one deep geological disposal site for long-lived waste is already in operation: The waste isolation pilot plant (WIPP) near Carlsbad, New Mexico, was licensed in 1998, but it can receive only waste from the US defense program. WIPP is embedded in a deep salt bed.

As of May 2014, only Finland and Sweden have selected the sites of their geological repositories. France should follow suite very soon. In the USA, a good deal of effort and money has been spent to build a repository (or long term storage facility?) in Yucca Mountain, Nevada, but President Obama has canceled the project and the USA is back to square one.

Before deciding to actually build an underground repository, a few countries have established deep underground laboratories to fully characterize the geological stratum considered for the future disposal site and are testing *in situ* the design and the handling systems.

2.3 *Decommissioning*[7]

At the end of their operating life, nuclear facilities are shut down, dismantled and then decommissioned. Decommissioning is an administrative step that releases the dismantled facility, totally or partially, from regulatory oversight. Dismantling consists of taking down production equipment, workshops and structures where nuclear materials were present, and then disposing of the resulting radioactive waste.

The IAEA defines three stages of dismantling:

- Stage 1: Removal from the facility of all radioactive materials and heat transport fluids. The locked-up facility remains under surveillance, maintenance, and monitoring.
- Stage 2: The reactor is significantly decontaminated, and remaining areas with important residual radioactivity levels are sealed. Radioactive components that can be easily dismantled are removed. Some surveillance and monitoring are maintained.
- Stage 3: All radioactivity above acceptable levels is removed. No further inspection, surveillance or monitoring is required.

In most cases, the bulk of the scrap metal or other materials is neither radioactive nor contaminated and can be recycled or disposed of along with conventional waste. Radioactive materials are sorted and packaged and sent to low level waste disposal sites. After decommissioning is complete, the land can be returned to unlimited use.

Decommissioning can be done very soon after the definitive shutdown of a facility, or it can be delayed for decades. Here are a few factors favoring immediate dismantlement:

- Decommissioning funds are available and costs are known.
- Low-level waste disposal sites are available.
- Experienced facility personnel and proven technologies are available.
- Presents a positive public perception.
- Eliminates corporate liability sooner and makes the site available for re-use.

On the other hand, the following factors may favor delayed dismantling:

- Smaller radioactive waste volumes.
- Lower staff radiation exposures.
- More time to resolve waste management issues.
- Benefit from technology enhancements.

The availability of disposal "routes" is especially important. For example, the lack of disposal sites licensed to accept graphite has delayed the decommissioning of the old French gas-cooled reactors.

2.4 *The Oklo phenomenon*[8]

In June 1972, at the Pierrelatte French enrichment plant devoted to Defense Applications, a routine mass spectrometry analysis of UF_6 feed material exhibited a discrepancy: there was only 0.7171% of ^{235}U in the uranium samples, instead of the magic 0.7202! Even though the discrepancy was small, it was so unusual that the French Atomic Energy Commission, CEA, operator of the plant, started a thorough investigation. First, it was determined that the discrepancy was not an artifact: the anomaly was confirmed in several measurements on other samples. Accidental contamination by depleted uranium from the plant itself was then eliminated and so was the use of reprocessed uranium as there was no ^{236}U in the samples. The investigators then traced the anomaly back through all the stages of uranium processing. Very soon it became clear that all the anomalous ore came from the very rich Oklo deposit in Gabon. In some shipments, the level of ^{235}U was as low as 0.44%. Between 1970 and 1972, in the 700 tons of uranium delivered from Oklo, the deficit of ^{235}U exceeded 200 kg, hardly a trifle!

The Oklo mine uranium was indeed different from natural uranium everywhere else. Why?

As early as August 1972, the hypothesis of very ancient fission chain reactions was formulated, and investigators started to search for fission products (or, rather, the granddaughters of hypothetical fission products). The spectrum of fission products is so distinctive that it constitutes an unmistakable marker that fission reactions have taken place. The presence of such fission products was clearly identified. *At some point in the uranium deposit history, it had become a "natural" nuclear reactor.* Later on, it was found that there were actually 15 reactor sites in Oklo, and another one in Bangombé, 30 km away from the main deposit. The discovery was duly heralded but many questions remained. When did the reactor "start"? How long did it "operate"? How was it "controlled"? The detective story was still not finished.

It was only around 2.2 billion years ago, that the patient work of photosynthesis accomplished by the first algae released enough oxygen in the earth's atmosphere for the surface water and ground water to become oxidizing. Only then could the uranium diluted in granite be leached out and concentrated before mineralization in places where oxidation-reduction would occur. Rich deposits cannot be older. On the other hand, for the past 1.5 billion years, ^{235}U has decayed sufficiently that its abundance is below a level which makes fission workable. It required many studies, in geology, chemistry and reactor physics to narrow the bracket of time to the present estimated value: the reactions must have started **1,950 $\pm$ 30 million years ago**.

The deposits were located several km deep[f], in very porous sandstone where the ground water concentration may have been as high as 40%. During the reactors' operation, the water temperature rose significantly, dissolving the silica and, by difference, increasing the concentration of uranium, therefore compensating for its depletion by fission. Furthermore, losing its silica, the surrounding sandstone became clay and thus prevented an excessive migration of groundwater and keeping the uranium in place. Physicists calculated that, varying from one zone to another, reactions did take place during an enormous length of time ranging **from 150,000 to 850,000 years**!

Even though significant alteration occurred in recent times when the tectonic uprising and erosion brought the reactors close to the surface, and especially when the Okolo Néné River gouged the valley, *the heavy*

---

[f]At such depth, the conditions of pressure and temperature were close to those of the PWR of today (350–400°C, 15–25 Mpa).

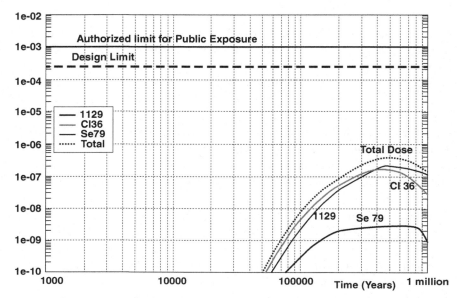

Fig. 5. HLW Deep Repository in Clay: Dose at the outlet in Sievert/year as a function of time.

elements thorium, uranium, and plutonium did not move at all, nor did the rare earth fission products, as well as zirconium, ruthenium, palladium, rhodium, and a few others. On the other hand, krypton, xenon, iodine, barium, and strontium have moved, but probably only after a few million years.

Soon after the discovery, and beyond the pure scientific thrill, the nuclear community was very excited by its implications, notably as a "natural analog" for the geologic disposal of high level radioactive waste (HLW). In a deep geological repository, one must contain the HLW from the spent nuclear fuel for around 500,000 years (Fig. 5). There, in Oklo, Mother Nature had contained *precisely the same radioactive elements* not for hundreds of thousands, not for millions, but for a couple of billion years, and without engineered barriers or special packaging. It is, especially true for the heavier elements which constitute most of the radio-toxicity of the HLW packages.

At the very least, this gives some confidence that the concept of deep geological repository is sound.

References

1. INES User's Manual (2009). NEA/IAEA, Vienna.
2. Three Mile Island Accident (2009). USNRC Backgrounder. August.
3. INSAG-7 (1992). The Chernobyl accident. Safety series No. 75, IAEA .
4. The Fukushima Nuclear Accident Independent Investigation Commission. The National Diet of Japan, July 2012.
5. UNSCEAR Report to the UN General Assembly, 2013.
6. International Experiences in Safety Cases for Geological Repositories (INTESC), OECD/NEA 6251, 2009.
7. Decommissioning of Nuclear Facilities. OECD/NEA factsheet, 2009.
8. Primitive nuclear reactors, Accessed on November 2015 at http://www.world-nuclear.org/info/Nuclear-Fuel-Cycle/Power-Reactors/Nuclear-Power-Reactors/.

Chapter 4

Indirect-Drive Inertial Confinement Fusion[*]

Erik Storm[†] and John D. Lindl[‡]

Lawrence Livermore National Laboratory
P.O. Box 808, Livermore, CA 94551
[†] *storm1@llnl.gov*
[‡] *lindl1@llnl.gov*

Laser driven inertial confinement fusion (ICF), an approach to thermo-nuclear fusion using mm-size fuel capsules filled with deuterium and tritium (DT) imploded either directly or indirectly by high power lasers, has been the subject of widespread theoretical and experimental studies since the early 1970s. The National Ignition Facility (NIF) at Lawrence Livermore National Laboratory (LLNL) supports a broad range of users in ignition science, other high-energy-density stockpile science, national security applications, and fundamental science and operates as an essential element of the National Nuclear Security Administration Stockpile Stewardship Program (SSP). The facility has achieved its design goal of 1.8 MJ and 500 TW of 3ω light on target. Recent results at NIF with the laser indirect-drive approach, have demonstrated a fusion yield of 26 kJ, more than double the energy invested in the DT fuel during the implosion. This yield is also more than twice the yield that could be obtained from compression heating alone and is the first time in a laboratory setting that the fusion yield is doubled as a result of self-heating by ^{4}He particles, a byproduct of the DT fusion reaction. Current experiments on the NIF are working to identify solutions to the remaining challenges on the path to ignition and propagating thermonuclear burn needed to produce net energy gain at NIF's laser energy of $\sim$2-MJ. Successful demonstration of ignition and gain will open the door

---

[*]This work was performed under the auspices of the U.S Department of Energy by Lawrence Livermore National Laboratory under Contract DE-AC52-07NA27344.

to new regimes in the laboratory in support of the SSP, but is also generally considered to be the crucial advance needed to consider ICF as an option for fusion power plants. This chapter summarizes the basic requirements of ICF and progress towards ignition on NIF, followed by requirements that would have to be considered for a program to develop fusion energy.

1 Introduction

Fusion is an attractive but challenging energy option for the future.[1-5] ICF uses drivers that compress energy in space and time to rapidly implode a capsule containing an equimolar mixture of DT. For spherical targets as the capsule radius decreases and the DT density and temperature increase, DT fusion reactions are initiated in a central region (hot spot) in the center of the compressed capsule that can reach pressures in excess of a hundred Gbars. These DT fusion reactions generate both 3.5 MeV ^{4}He (alpha) particles, whose range is short enough that they can be stopped in the compressed fuel, and 14.1-MeV neutrons, whose range is long enough that most escape the compressed fuel. Under suitable conditions, a fusion burn front propagating from this hot spot into surrounding cold fuel is capable of generating significant energy gain (fusion yield/driver input energy >1).[6,7] The process of ignition and burn in the compressed fuel takes 10's–100's of picosecond. In contrast, magnetic fusion energy (MFE) uses powerful magnetic fields to confine a low density DT plasma and to generate the conditions required to sustain the burning plasma for a sufficiently long time (seconds) to generate energy gain. In contrast to ICF, plasma pressures, limited by the strength of the magnetic fields are only a few atmospheres.

The roadmap for transforming fusion energy into a source of electricity can be thought of as a five-step process: (1) demonstrate understanding of the underlying fundamental physics principles, (2) demonstrate ignition and net energy gain, (3) execute an R&D program to develop the technologies required for transforming the fusion products into useful energy (4) build and demonstrate a prototype power plant, and (5) build commercial power plants. The advances made over the last 50 years means that major progress has been made on Step 1 for both MFE[5] and ICF.[1,8,9] A research program for Step 2 for ICF is in place and will hopefully be achieved in the near future (see Section 3), and progress has already been made on many of the technologies for Step 3 (see Section 5).

The current experiments designed to demonstrate ignition and net energy gain in ICF are using a central hot spot ignition (HSI) target in an indirect-drive configuration.[9] HSI relies on simultaneous compression and ignition of the spherical DT-filled capsule. In the laser indirect-drive configuration, the capsule is placed inside a cavity (most research today has focused on cylindrical cavities) of a high-Z metal (a hohlraum). Focusing the laser energy onto the interior walls of the hohlraum and converting it to X-rays provide the energy source needed to implode the capsule. The X-rays are absorbed on the surface of the capsule, which implodes, reacting like a spherical rocket. The small (a few % of the total DT fuel mass), high-temperature central part of the imploded fuel provides the "spark," which ignites the cold, high-density portion of the surrounding fuel. The scientific basis for indirectly driven HSI targets has been intensively developed since the early 1970s, and a research program to demonstrate ignition and net energy gain with indirectly driven HSI targets using the National Ignition Facility (NIF) in support of the SSP is now underway.

Laser driven ICF research is thus at a key juncture. The demonstration of fusion ignition and energy gain in an experimental setting would provide a motivation and the basis for the transition from scientific research to considering Inertial Fusion Energy (IFE) as an option for commercial power plants.

Section 2 of this paper reviews the basic requirements of ICF, Section 3 provides a review of progress toward the demonstration of ignition using indirect-drive targets on the NIF, Section 4 reviews the basic requirements for the development of fusion energy using ICF, Section 5 reviews progress in the development of key technologies that would be required for fusion energy using indirect-drive ICF targets, and Section 6 provides concluding remarks.

2 The Physics of ICF

2.1 *Review of Basic Concepts*

The basic idea of ICF is for a driver to deliver sufficient energy, power, and intensity to a capsule containing DT to make the capsule implode. The implosion is designed to compress and heat the DT mixture to a temperature at which thermonuclear burn is ignited. The self-sustaining thermonuclear reaction releases the energy of the DT mixture, and if the energy output is greater than the driver energy supplied, the gain is >1. To cause the $D + T \rightarrow {}^4He + n + 17.6$ MeV reaction to take place to a reasonable degree, the fuel must be raised to a temperature of 5–10 keV or greater,[a]

---

[a]Energy and temperature are related by Boltzmann's constant, k = 1.3807 × 10⁻²³ J/K. Therefore 10 keV is equivalent to 1.16×10^8 K i.e. about 116 million degrees Kelvin.

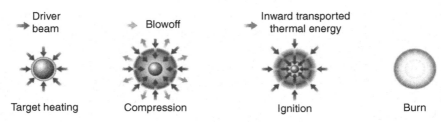

Fig. 1. Basic sequence of events for high gain ICF.

and for the reaction to be self-sustaining there must be a sufficiently high particle density. Densities of interest to ICF are in the range of 200–2000 g/cm^3 and pressures in the range of 100–500 Gbars, depending on the target and performance required. (Reference 7 extensively discusses the details of all aspects of ICF physics.)

The sequence of events leading to DT burn is shown in Fig. 1. The driver deposits its energy either by electron conduction (with the driver — lasers or particle beams directly focused on the capsule) or indirectly in the outer layers of the fuel capsule (via X-rays generated by converting the driver energy to X-rays in a high-Z cavity surrounding the capsule). This energy deposition in turn ablates the surface of the fuel capsule, and the ablation acts as a rocket exhaust to drive the implosion of the capsule.

The ablation generates pressures which are precisely tailored in time to rise from 1 to >100 Mbar over a period of about $10-30 \times 10^{-9}$ sec and accelerates the capsule inward to velocities of $3-4 \times 10^7$ cm/sec ($>10^{-3}$ of light speed!). The acceleration continues until the internal pressures exceed the ablation pressure. At that time, the rapidly converging shell begins to decelerate and compresses the central portion of the fuel. In its final configuration, the fuel is nearly isobaric at pressures of ~100–500 Gbars (depending on the capsule design) but consists of two very different regions — a central hot spot containing 2–10% of the fuel and a dense main fuel region with a peak density of ~1000 g/cm^3. Fusion initiates in the central hot spot, and a thermonuclear burn front propagates through the dense, main fuel producing high-energy gain. At these conditions, approximately 30% of the fuel undergoes fusion. To achieve these conditions, high gain ICF targets have features and requirements similar to those shown in Fig. 2.

This rather straightforward sequence of events is complicated by the impracticality of building drivers of arbitrarily high energy, power and focusability, and by some basic facts of nature. The basic facts of nature are: (1) the implosion must be carried out in a manner that results in

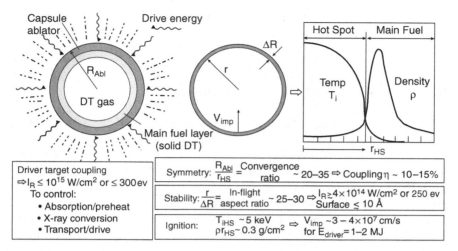

Fig. 2. Target physics specifications and requirements for high gain ICF.[6,7]

little increase in fuel entropy during the implosion or else the required fuel compression will put unacceptable demands on driver energy, (2) the energy from the driver must be delivered to the capsule with a very high degree of spherical uniformity, and (3) potential hydrodynamic instabilities demand a specific type of ablative implosion to avoid mixing of the outer shell with the fuel and subsequent degradation of the burn efficiency.

The underlying questions with achieving high gain ICF have always been: can we make an appropriately configured fuel capsule; can we produce a sufficiently uniform pressure pulse to cause the 30- to 40-fold radial convergence of the capsule required to achieve the necessary high compression; and do we fully understand the driver/target interaction physics, the hydrodynamics, and the physics of DT burn on which our models are based? A few simple arguments will indicate the design constraints for a successful high gain target.

2.2 *DT Burn Physics*

The thermonuclear burn rate for a DT plasma can be written

$$\mathrm{dn/dt} = n_D\, n_T \langle \sigma v \rangle, \qquad (1)$$

where $\langle \sigma v \rangle$ is the Maxwellian averaged cross section and $n_D(t)$ and $n_T(t)$ are the instantaneous deuteron and triton number densities. We define the fuel burn up fraction as $\phi = n/n_o$, where n_o is the initial total fuel number density. If we assume a nearly constant burn rate, then for an equimolar mixture of D and T we can integrate (1) to find

$$\phi/(1 - \phi) = n_o \tau / 2 \langle \sigma v \rangle. \tag{2}$$

For DT at about 20 keV, we have approximately

$$\phi = n_o \tau / (n_o \tau + 5 \times 10^{15}) = \rho r / (\rho r + 6), \tag{3}$$

where n has units of particles per cubic centimeter and τ has units of seconds. In the equality on the right, ρr has units of g/cm^2 and is obtained from the formula containing $n\tau$ by setting $n = 2.4 \times 10^{23} \rho$ for DT and $\tau = r/3v_s$, where r is the radius of the compressed fuel and v_s is the sound speed. We use $r/3$ in this rough estimate of the confinement time because most of the mass is located within the outer 1/3 of the compressed fuel. Once most of the compressed fuel drops substantially in density, the burn rate falls rapidly.

This formula for ϕ for ICF agrees well with detailed burn-up simulations of most high gain ICF targets. One also notes that in order to achieve reasonable burn-up, say between 1/4 and 1/3, the fuel ρr should be between 2 and 3 g/cm^2.

Ignition of the DT fuel occurs when the alpha particle energy deposition into the fuel plasma during one energy confinement time equals the energy required to heat the plasma. If the electron and ion temperatures are equal, then the energy required for heating the fuel with 100% efficiency equals

$$E_{\text{DT heating}} = 0.115 \times 10^9 \text{ T (J/g)} = 2.3 \times 10^9 \text{(J/g)},$$
$$\text{at 20 keV with } T_{\text{ion}} = T_e. \tag{4}$$

A DT plasma produces 3.34×10^{11} J/g if it burns completely. The product of the burn efficiency and this specific energy gives the specific thermonuclear energy produced. About 20% of the energy produced in the DT reaction is in 3.5 MeV alpha particles and the rest in 14.1 MeV neutrons. The alpha particles have a range of about 0.3 g/cm^2 in a 10 keV plasma. The central hot spot in an igniting ICF target has a ρr sufficient to capture most of the alpha energy, providing the energy for a self-sustaining burn wave that propagates from the hot spot to the surrounding cold fuel. The 14.1 MeV neutrons have a scattering range in DT of about 5 g/cm^2. Typical ICF implosions have a total compressed fuel $\rho r \sim 2$–3 g/cm^2, so most of the neutrons escape without contributing their energy to the compressed fuel.

Using just the energy from the alpha particle, the specific thermonuclear energy available to heat the plasma is given by

$$E_{\text{thermonuclear } \alpha \text{ particle}} = 6.68 \times 10^{10} n\tau / (n\tau + 5 \times 10^{15}) \text{ (J/g) at 20 keV.}$$

(5)

Setting these two equal in this simple model, the requirement for ignition is $n\tau > 1.7 \times 10^{14}$ sec/cm^3. This is known as the Lawson Criterion, and corresponds to $\phi = 0.034$, which is adequate for MFE if losses sustained in maintaining the magnetic field are negligible.

2.3 *Compression and Central Ignition*

For ICF this is equivalent to a $\rho r \sim 0.3$ g/cm^2 at 20 keV. However, if the fuel is surrounded by a denser region of DT of $\rho r \sim 2$–3 g/cm^2, the ideal ignition condition can be relaxed to $\rho r \sim 0.3$ g/cm^2 at ~ 5 keV. The increased confinement time then allows the alpha deposition to bring the fuel to ignition temperature (20 keV) before decompression. For ICF this is also called marginal ignition. However, ignition is inadequate for all approaches to ICF. For radiation driven concepts using lasers, one must first overcome a factor of 5–10 in hohlraum efficiency and a factor of 5–10 in capsule implosion efficiency. For IFE power generation, we must also account for another factor of 5–10 in driver efficiency. ICF is able to overcome these efficiency factors by utilizing two effects: compression and central, or hot spot ignition.

More than an order of magnitude in efficiency comes from **Compression**. If a ρr of 2.5 g/cm^2 can be achieved, the burn efficiency rises to 30%. This ρr is more than sufficient to absorb the 3.5-MeV alpha particles, while most of the 14.1-MeV neutrons escape without contributing a significant fraction of their energy to the fuel (their range is about 5 g/cm^2 in DT). However, if the DT is uncompressed, a ρr of 2.5 g/cm^2 requires a sphere of solid DT almost 25 cm in diameter with a mass of ~ 1.5 kg. Not only would such a large fuel mass produce an unacceptably large yield ($\sim 1.5 \times 10^{14}$ J, or ~ 35 kT) but it would take about 10^{13} J to heat it to ignition. Since mass m $\sim (\rho r)^3 / \rho^2$ for a uniform density sphere, the fuel mass and hence the driver energy required (which is proportional to the fuel mass) for a given ρr can be reduced by a large factor by compression.

The kinetic energy of the imploding fuel shell must compress the fuel to the required density and then heat it to ignition temperature (5–10 keV). If the fuel is compressed to a density of 500 g/cm^3, ($\sim 2{,}000$ times liquid density) the mass required for a ρr of 2.5 g/cm^2 drops to ~ 2 mg for a spherical DT shell of radius R and shell thickness $\Delta R = R/2$ (a simple approximation to the radial density distribution shown in Fig. 2). The fusion yield for a burn-up fraction of ∼30% now drops to a manageable 200 MJ.

Compression is attractive from an energetic point of view, however, only if it is "cheap" to do compared to the energy required for heating the DT. If the compression can be done with the fuel in a near Fermi-degenerate state (which is possible if the driver energy is delivered with a precise time history), this is indeed the case. The specific energy for compression of Fermi-degenerate DT is

$$\varepsilon_{FD}(J/g) = 3 \times 10^5 \rho^{2/3} \quad \text{(with } \rho \text{ in g/cm}^3\text{) or } \varepsilon_{FD} \sim 1\text{--}3 \times 10^7 \text{ J/g} \quad (6)$$

for DT in the 200–1000 g/cm^3 range. This is only a few percentage of the energy needed to heat DT to the 5–10 keV required for ignition. (The specific heat for DT per keV is ∼1.15 × 10^8 J/g.) While only ∼40 kJ is needed for the compression of 2 mg to an average density of 500 g/cm^3 and $\rho r = 2.5$ g/cm^2, an implosion energy of ∼1–2 MJ is required to heat the 2 mg to 5–10 keV.

If we assume a driver-to-implosion efficiency of 5%, the driver would have to deliver 20–40 MJ. While the yield of ∼200 MJ would be sufficient for IFE, the gain of 5–10 is inadequate and the cost of the driver would be prohibitive.

To overcome this difficulty and to provide more than another order of magnitude, ICF capsules use **Central Ignition** and propagation of the burn into compressed DT to achieve high gain. In a typical high gain capsule, only a few percent (∼2%) of the fuel has to be heated to ignition conditions. The energy invested to heat this part of the 2 mg of fuel, would only be ∼40 kJ, and alpha particles produced in this region ignite the rest of the surrounding highly compressed, cooler fuel. If this is done, the energy used to ignite the fuel becomes comparable to the compression energy. Since the main fuel is also heated to about 100 eV, the total kinetic required would be about 100 kJ.

An implosion velocity of ∼3.3 × 10^7 cm/sec is required to give the 2 mg of fuel this kinetic energy. This results in a laser of about 2 MJ, not an unreasonable requirement. With the fusion yield of 200 MJ this would produce a gain of about 100, more than sufficient for IFE. We note that when using an actual final compressed DT configuration such as shown in Fig. 2, the burn-up fraction, yields and resulting gains would be reduced, but gains of about 50, adequate for IFE (see Section 4) could still be achievable for a 2.5–3.5 MJ laser, depending on the outcome of physics issues still being investigated on NIF and elsewhere.

2.4 *Fluid Instabilities, Mix and Low Entropy Implosions*

Although energetically attractive, the physics of compression and hot spot ignition is very demanding. At ignition, the radius of the hot spot is a factor of 30–40 smaller than the initial radius of the capsule. This is achieved by forming the fuel into a spherical shell. This increases the effective volume over which PdV work can be done on the fuel and lowers the pressure required. With peak pressures of about 100 Mbar, convergence ratios of about 40 are required by this technique. To maintain a near-spherical implosion with this convergence ratio requires a drive flux uniform to about 1–2% over the surface of the capsule.

A second hurdle for successful ICF implosions is the effect of hydrodynamic instabilities. A description of the detailed analysis of the growth of Raleigh–Taylor instabilities (caused by the low-density ablation region pushing on the high-density imploding shell and the low density central region decelerating the dense fuel at stagnation) is outside the scope of this chapter. However, both analytical models and detailed numerical simulations show that the growth of instabilities is proportional to the ratio of the shell radius (R) to the shell thickness ΔR. Since the shell is compressed during the implosion, we must use a representative value of the shell thickness during the implosion, rather than its initial value, to get the so-called in-flight aspect ratio. To keep the imploding shell from breaking up and the resulting mix of cold ablator and cold fuel from mixing into the hot spot and quenching the ignition, this in-flight aspect ratio should be kept to a maximum of about 25–35. Detailed numerical simulations suggest that the surface of the ablator must also be smooth to about 1 nm and the DT ice layers to about 0.5 μm.

A third constraint, is the requirement that the entropy generated during the implosion be limited to about that generated by a 1-Mbar shock passing through the initially uncompressed layer of solid or liquid DT. This allows the compression process to keep the fuel as near as possible to a Fermi-degenerate state; i.e. as low an entropy generation and heating as possible. Keeping the first shock to about 1-Mbar and each succeeding shock to not more than 4 times the pressure of the preceding shock satisfies the low entropy generation requirement.

The peak ablation pressures (P_{abl}) required to generate the implosion velocities and the compression needed for high gain ICF implosions are around 100 Mbar.

The development of the technology to fabricate targets with the precision mentioned and the development of the experimental and computational

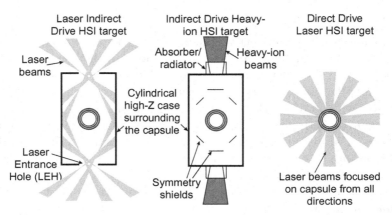

Fig. 3.   Schematic laser and heavy-ion indirect-drive and laser direct-drive HSI targets.

tools to accomplish implosions with this kind of accuracy has been one of the major achievements of the worldwide ICF programs over the past 40 years.

2.5 *Indirect and Direct Drive Approach to Hot Spot Ignition ICF*

The two fundamental approaches to laser driven ICF are differentiated by the mechanism chosen for the energy transport to drive the ablation and implosion (see Fig. 3).

In the **laser direct-drive** approach, laser beams are arrayed around the target in a near uniform pattern and focused directly on the fuel capsule. (The laser direct-drive approach to ICF and IFE is covered extensively in Chapter 5 of this book and in Refs. 10–21.) The energy is transported from the absorption region to the ablation surface by electron conduction. The absorption occurs at a particle density equal to or less than the critical plasma density n_c (cm^{-3}) $= 10^{21}/\lambda^2$, where λ is the laser wavelength in μm. Typically, particle densities at the ablation front are 2–3×10^{23} cm^{-3}, and electron conduction must bridge the gap between the two.

Two fundamental requirements for direct drive are efficient absorption of the laser light in the underdense plasma region by collisional processes (to avoid excessive laser plasma interaction (LPI) collective instabilities and the subsequent energy loss and generation of hot electrons and redistribution of the incident laser light through "seeded" Brillouin Scattering, Raman Scattering and Two Plasmon Decay) and a pressure distribution at the ablation surface uniform to <1–2%. Another issue for direct-drive results from high spatial frequency spatial modulation of the laser beams that produces "early time" mass defects ("imprinting") that serve as seeds for hydrodynamic instabilities. To avoid high levels of laser-plasma instabilities and generation of energetic electrons while producing the necessary ablation pressures, direct-drive targets typically employ laser wavelengths ≤ 0.35 μm. This increases the density at which the laser light is absorbed, which increases inverse bremsstrahlung collisional absorption. This, however, reduces the region between the absorption and ablation front, which reduces the potential smoothing by electron conduction. Uniformity must therefore be obtained by overlapping a large number of nearly ideal beams and the use of beam smoothing techniques.

Direct drive is a more efficient approach since the laser energy directly irradiates the capsule (although the rocket efficiency resulting from the higher plasma exhaust velocity is reduced about a factor of 2), and simulations project about 2 × higher gain than indirect-drive HSI, but the stringent illumination uniformity requirements tend to offset some of this potential advantage.

In **laser indirect-drive**, the laser energy is first converted into X-rays inside a high-Z enclosure (a hohlraum).[9] The fuel capsule is contained inside the hohlraum, and the X-rays are used to drive the ablation of the surface of the capsule, decoupling the absorption and transport process. The hohlraum smoothes small-scale non-uniformities in the driver beams, relaxing the requirements on laser beam uniformity although smoothing is still required to suppress LPI. A wide variety of possible laser geometries can minimize the remaining long wavelength asymmetry. The most studied, laser-driven hohlraums are cylinders with two laser entrance holes (LEH).

Target physics specifications for laser-driven indirect-drive targets were shown in Fig. 2. Driver-target coupling issues limit the laser intensities inside the hohlraums to $\leq 10^{15}$ W/cm^2 resulting in a radiation temperature limit of about 300 eV. This limitation is primarily the result of laser-driven plasma instabilities, which result in the scattering of laser light and the generation of high-energy electrons. Light scattering degrades symmetry and high-energy electrons cause capsule and fuel preheat which reduces the compression. With the convergence requirement of ~30–40, the X-ray flux on the capsule must be uniform to 1–2%. To achieve this level of uniformity, the hohlraum areas must typically be 15–25 times larger than the capsule surface area. This large case-to-capsule area limits the coupling efficiency of the driver energy to the capsule to 10–15%, although future research may ultimately show it to be possible to increase this to about 20% by optimizing hohlraum and LEH geometries with respect to capsule surface area (and possibly materials and hohlraum composition).

Relative to direct drive, radiation or indirect-drive targets are less sensitive to the effects of hydrodynamic instability during the implosion. However, since it is a two-step process (absorbed laser light is first converted to X-rays and the capsule only absorbs a fraction of these X-rays), indirect-drive targets are less efficient. The better-matched plasma exhaust velocity of X-ray driven ablation also results in improved rocket efficiency.

Heavy-ion indirect-drive. Capsule performance is independent of the source of the X-rays as long as they have the appropriate spectral, temporal, and spatial characteristics. This means that the X-rays to drive the capsule can also be produced by other drivers such as heavy ion (HI) beams.[22,23] The hohlraum would still be a cylindrically shaped high-z case, but the ion beams would be focused on absorbers whose locations are chosen based on the same basic principles as those for a laser-driven target. The ion beam energy is then reradiated in the form of X-rays. A typical HI beam indirect-drive target is shown in Fig. 3. Unfortunately, limited research is ongoing today to evaluate this driver option.

Pulse shaping. Reaching a peak drive pressure of some 100 Mbar coupled with the condition for low entropy generation means that four or more shocks, each one not more than a factor of 4 greater than the preceding one, must be used to reach the peak pressure. The scheme is to send the shocks through the solid (or liquid) DT fuel in such a way that they all coalesce near the interior edge of the cryogenic fuel layer and there combined with the initial gas fill, produce the hot spot required for ignition.

Figure 4 shows the X-ray drive pulse for an indirectly driven target. The laser pulse required to produce this X-ray drive pulse is also shown. A laser pulse with a shape similar to the X-ray drive pulse of Fig. 4 would be required for a laser direct-drive HSI target.

Wavelength dependence. Nearly all of the experimental data for ICF targets over the past 40 years have been obtained with solid-state lasers operating at 1.06 μm or frequency converted to 0.53, 0.35 or 0.25 μm, (Table 1) although some work has also been done with krypton fluoride (KrF) gas lasers at 0.25 μm at the Naval Research Laboratory[19−21] and, primarily in Russia, with iodine-based gas lasers at 1.315 μm wavelength.[24]

The coupling of laser energy to the target, both for the indirect-drive and the direct-drive approach is greatly enhanced at shorter wavelengths since the shorter the laser wavelength, the higher the absorption of the laser energy and the smaller the conversion of laser energy to high-energy

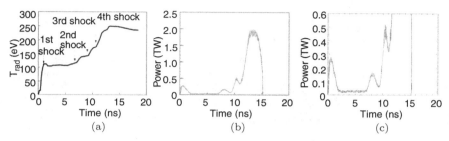

Fig. 4. (a) Capsule X-ray drive pulse for indirectly driven HSI target: (b) and (c) the required laser power from 1 of 192 NIF beams as a function of time. (b) and (c) show an overlap of data from 16 shots, demonstrating the repeatability and precision of NIF.

electrons (which would preheat the shell and DT fuel and cause performance degradation) and other deleterious LPI effects.

High gain direct-drive HSI target designs generally assume laser light with a wavelength 0.25–0.35 μm. Indirectly driven HSI targets with a $v_{\mathrm{imp}} = 3.5-4 \times 10^7$ cm/sec generated by hohlraum temperatures >250 eV generally require laser wavelengths of 0.35 μm. Higher yield indirect-drive HSI targets that can operate at temperatures <250 eV may tolerate 0.53 μm although LPI with these large homogeneous plasmas is a concern that would have to be addressed experimentally given the complex nonlinear physics that may be present.

2.6 *Alternative Ignition Concepts*

For conventional HSI targets, the driver pulse (laser or heavy ion) assembles the fuel at high density and imparts a sufficiently high velocity ($\sim 3.5 \times 10^7$ cm/sec) to the imploding shell so that its PdV work create both the high main fuel density and the central ignition hot spot on stagnation.

By contrast, there are target options, such as fast ignition[25–31] and shock ignition[32–37], which decouple the fuel assembly and ignition phases. For both, the cryogenic DT shell is first compressed to a $\rho r \sim 2-3$ g/cm^2 at a density of $\sim 300-500$ g/cm^3 at a low velocity ($\sim 2 \times 10^7$ cm/sec) using a driver of lower peak power and lower total energy. For a shock ignition target, the assembled fuel is then separately ignited from a central hot spot heated by a strong, spherically convergent shock driven by a high intensity spike at the end of the compression pulse. For the fast ignition target, an extremely high peak intensity ($\sim 10^{20}$ W/cm^2) short pulse (10–20 ps) laser focused to a small ($\sim 20-30$ μm diameter) spot at the edge of the high density compressed fuel provides a pulse of MeV electrons (or ions) to

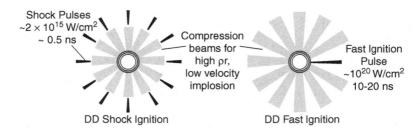

Fig. 5.   Schematic of directly driven shock ignition and fast ignition targets.

initiate ignition and propagating burn which spreads through the remainder of the fuel (Fig. 5).

The majority of the laser energy is contained in the main, compression portion of the pulse, with ~20% for the separate ignition pulse. Because the implosion velocity is less that required for conventional HSI, more fuel mass can be assembled for the same kinetic energy in the shell, offering higher fusion yields for the same laser energy. For the fast ignition scheme the compression could be achieved either by direct or indirect-drive, although the overall gain will be slightly lower with indirect-drive. For shock ignition, because of symmetry requirements, it does not appear possible to couple indirect-drive for the initial compression phase with the direct-drive pulse needed for the strong final shock. There are many challenges to both of these approaches and a discussion and summary of the focused research is beyond the scope of this chapter.

2.7 *Summary of ICF Target Performance*

Figure 6 shows calculated potential target performance for HSI targets for both indirect and direct-drive with lasers and HI, and direct-drive shock and fast ignitions.

In contrast to directly driven HSI targets, for which KrF lasers with a wavelength of 0.25 μm were assumed in calculations of these gain curves, simulations show that directly driven shock and fast ignition targets can achieve adequate symmetry and convergence with 0.35 μm laser wavelength, available with frequency converted solid-state lasers. Although all of these gain curves will have significant levels of uncertainty until ignition is achieved in the laboratory, the gain curves for the advanced less developed concepts have more uncertainty than gain curves for hot spot ignition.

As noted above, the projected gain for various ICF target concepts is based on widely varying levels of knowledge and assessment. In particular,

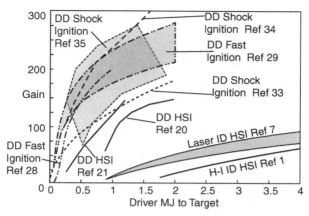

Fig. 6. Target gain for ICF targets and drivers. Potential gains for all of these ICF approaches are based on the judgment of scientists working in the field and are subject to change as more information becomes available. The basis behind the laser indirect-drive curves is discussed in more detail below.

fast ignition and shock ignition targets are at an early stage of investigation, with complex physics and engineering issues still to solve, and it is not clear what level of the postulated high gain performance, if any, will remain after these issues have been resolved, nor how long such an R&D effort would take. Finally, as will be discussed in Section 4, the gain of an ICF target is not the only metric from which to choose an IFE system.

3 Progress towards Ignition with Laser Indirect-Drive ICF

3.1 *Research — The first 40 Years*

The physics of ICF has been the subject of extensive theoretical and experimental studies around the world for more than 40 years. The principal institutions currently doing experiments with indirectly and directly driven targets are listed in Table 1.[11,12,19,38−53] Table 2 shows the main future facilities under construction with anticipated completion dates.[54−58]

Early efforts to explore the physics of ICF were carried out in the 1970's and early 1980's in the U.S. with Nd:Glass lasers operating at 1.06 μm with the Chroma laser at KMSF in Ann Arbor[59], with the Janus, Cyclops, Argus and Shiva lasers at LLNL[60−63] and the Helios CO_2 laser operating at 10.6 μm at LANL.[64] Work with Nd:Glass lasers operating at 1.06 μm was also carried out at the CEA-CEL[65] and the Ecole Polytechnique[66] in France, the Lebdev and Kurchatov Institutes in the former USSR, and at the University of Osaka in Japan. (See Ref. 67 for a discussion of early ICF work at these

Table 1:   Main Institutions Engaged in ICF R&D.[2]

Institute-location	Facility: Laser type*	Laser parameters	Main R&D
LLE, U Rochester, US	Omega/Omega EP	40 kJ/2 ns; 2.5kJ/10 ps	1; 2; 3; 6; 7
ILE, Osaka Univ, Japan	Gekko XII; FIREX-I	2–5 kJ/2ns- 500J/1 ps	1; 3; 6; 7
Rutherford Labs, UK	Vulcan	2.6 kJ/1ns; 500J/0.5ps	3; 5; 6; 7
Ecole Polytech., France	LULI-2000; Pico2000	2kJ/1.5ns; 200J/1ps	1; 5; 6; 7
NRL, Wash. D.C. US	Nike, KrF	5 kJ/4ns	1
LANL, Los Alamos, US	Trident	100J/500fs	6, 7
LLNL, Livermore, US	NIF	1.3–1.8 MJ/10–20 ns	Ignition;1; 2; 7

[2]* All lasers except Nike are Nd:glass; 1 = Direct-Drive HSI; 2 = Indirect-Drive HSI; 3 = Direct-Drive Fast Ignition; 4 = Direct Drive Shock Ignition; 5 = Fast Ignition Physics; 6 = High Field Physics ($I = 10^{20}$ W/cm^2); 7 = Basic High Energy Density Physics.

Table 2:   ICF Facilities under Construction.

Institute-location	Facility: Laser type*	Laser parameters	Main R&D
CEA/CESTA, France	LMJ (~2025)	1.1–1.5 MJ/10–20 ns	Ignition; 2; 7
CEA/CESTA, France	PETAL (~2017)	3.5 kJ/5 ps	5; 6
LFRC, China	SG III (2016)	180 kJ, 3 ns	2; 7
LFRC, China	SG IV (~2020–2022)	1.1-1.5 MJ/10–20 ns	Ignition; 2; 7

latter facilities.) Results from these facilities made it clear that to avoid high levels of laser-plasma instabilities and the deleterious effects resulting from generation of very energetic electrons, laser wavelengths less than ~0.5 μm would be required.

From the mid 1980s through the 1990s indirect-drive ICF was extensively studied with the frequency converted Nd:Glass lasers Nova at LLNL, and Phébus at the CEA laboratory in France. By the early 1990s the progress in the understanding of high energy density and laser plasma interaction physics, the development of experimental and computational tools of ICF, and the mission need arising from the expected cessation of nuclear weapons testing was sufficient to lead the DOE in the US and the CEA in France to begin construction of NIF and LMJ. (See Ref. 8 for a summary of these results.)

The main effort for DD HSI over the past decade has been conducted at LLE on their Omega facility and at NRL with the Nike KrF laser[10−21] (See Chapter 5). Ancillary experiments have also been carried out at ILE in Japan,[38−40] Rutherford Laboratory in the UK,[41,42] Ecole Polytechnique in France[43] and at LANL[44,45] in the U.S. Starting in 2012, DD HSI target experiments have also been performed on NIF. DD HSI on NIF utilizes the present NIF illumination geometry that is designed for indirect drive ("polar Direct Drive").

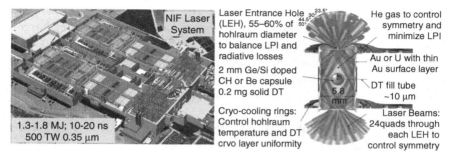

Fig. 7. NIF and the indirect-drive point design for the national ignition campaign.

3.2 *Laser Indirect-Drive Ignition Studies on the NIF*

Given the large experimental data base for indirect-drive HSI as well as the reality that the only experimental facility in Table 1 capable of exploring ICF ignition and net energy gain within this decade is NIF, the first potential demonstration ignition will therefore be done with laser indirect-drive HSI targets on NIF.[46–53] (References 46–53 provide detailed descriptions of the NIF and the development of the ICF physics and technology basis leading to NIF.)

Figure 7 shows a cut-away of NIF with the two laser bays, each with 96 beams and two switchyards, which rearrange the 192 beams into two sets of 24 groups of 4 beams entering into the target chamber from the top and bottom. NIF is the world's largest and most energetic laser facility for research in ICF and high-energy-density stockpile science. NIF was designed and built as an essential element of the NNSA Stockpile Stewardship Program.[51]

NIF also performs experiments for fundamental science, other national security missions, and the potential use of ICF as a source of renewable energy. The groundwork for NIF began after the 1990 National Academy of Science Review of the ICF Program. This work included the Nova Technical Contract,[49–51] with a series of physics objectives to be demonstrated on the Nova laser, and the operation of a scientific prototype of a NIF beam called Beamlet.[52,53] NIF has demonstrated that it can meet the laser performance goals for the National Ignition Campaign (NIC) indirect-drive point design by simultaneously meeting the requirements for temporal pulse shaping, focal-spot conditioning, peak power of 500 TW and total energy of up to 2 MJ for representative ignition pulses at 0.35 μm.

The NIC, established by the NNSA in 2005, was responsible for transitioning the NIF from a construction project to a national user facility.

Besides the operation and optimization of the use of the NIF laser, the NIC program was responsible for developing capabilities including target fabrication facilities; cryogenic layering capabilities; over 60 optical, X-ray, and nuclear diagnostic systems; experimental platforms; and a wide range of other NIF facility infrastructure. The goal of the ignition physics program on NIF is to demonstrate an ignition platform with laser indirect-drive HSI targets.

At the end of the NIC in 2012, implosion parameters had been demonstrated at ∼80–90% of most ignition point design values, though not simultaneously.[8,68] In low-mix implosions the nuclear yield was a factor of 3–10 × the simulated values. The yields were also a factor of 3–10 below that at which the yield from the additional heating by alpha particles would dominate the yield from compression alone. Since the hot spot temperature and size were reasonably close to predicted values, the principal reason for the reduced yield appeared to be a hot spot density and pressure that was factor of ∼2–3 below the simulated values. Ablator mix into the hot spot was observed at lower velocities than predicted[8,71] and correlated strongly with the measured ion temperature and yield. Indications were that this mix was being driven by hydrodynamic instability, Rayleigh–Taylor and Richtmyer–Meshkov (RM), at the ablation front. Seeds for the growth of these instabilities include capsule surface roughness and the capsule "tent" support structure. Several diagnostic measurements — both in-flight and at fuel stagnation — revealed time-dependent low-mode shape asymmetries in the ablator and cold fuel which may explain some of the deficit in pressure. These low mode asymmetries result in inefficient conversion of imploding fuel kinetic energy into fuel compression, as well as generating distorted hot spots with more surface area and larger thermal conduction losses from the hot spot to the surrounding fuel. Diagnostic measurements also indicated larger than expected mix of ablator material into the hot spot. In the most detailed simulations of the NIC implosions, the growth of perturbations from the tent support structure, and low mode time dependent radiation flux asymmetry are the largest sources of degradation from 1D performance.

In order to achieve the required symmetric implosion shape, the NIF experiments typically relied on a LPI process called Cross-Beam Energy Transfer (CBET)[69,70] to transfer a large fraction (∼30%) of the energy and power from the outer to the inner beams via an ion acoustic wave. This process is sensitive to the laser intensity and the plasma conditions in the region of the overlapping beams, and is not yet well modeled in an integrated fashion in simulation codes. The time and spatially varying CBET makes it difficult to adequately control the time dependent symmetry of the imploding capsule and has consequently contributed to the decrease in implosion performance as the laser power and capsule implosion velocity were increased. See Refs. 8 and 68–75 for an in-depth discussion of the results obtained during the 3-year duration of the NIC.

Following the completion of NIC in 2012, the theoretical and experimental efforts have been focused on understanding the underlying physics issues responsible for the deviation from modeled performance, with particular emphasis on the areas of low-mode shape asymmetry and mix. Below we discuss the results of implosions of CH (plastic) capsules with high-foot drive conditions that are more robust to hydrodynamic mix; results of implosions in Rugby-shaped hohlraums with reduced CBET; and implosions with high density carbon (HDC) ablators in low density or near vacuum hohlraums where CBET is minimal.

3.3 *The High Foot Campaign*

The goal of the **'High-foot' Campaign** was to modify the drive to create a more one-dimensional and robust implosion that is more resistant to ablation driven hydrodynamic instabilities and the resulting mix of the ablator into the DT fuel.[76–83] Figure 8 shows several of the pulse shapes that have been used in NIF experiments. The laser power in terawatts (TW) is given as a function of time. As discussed earlier, the purpose of the laser pulse shape applied to the capsule is to generate a succession of impulses whose timing and pressure is designed to control the entropy of the fuel. A single very strong shock generated by a rapid turn on of the laser to peak power would result in very high fuel entropy and very poor compression. A succession of four shocks, properly timed, of increasing strength can leave the fuel in a state of near optimal compressibility. A measure of the entropy or compressibility is the fuel adiabat α, which is the ratio of pressure needed to compress the fuel to a given density to the Fermi-degenerate pressure, which is the minimum possible pressure arising from quantum mechanics. The DT fuel must remain almost Fermi degenerate in order to minimize the pressure required for a given compression. Maintaining adequately low entropy for the entire fuel volume requires that each shock be carefully controlled in strength and launch time.

The low-foot 4-shock pulse shown in Fig. 8 was used in the NIC campaign from 2009 to 2012. The 3-shock high-foot CH design uses the same capsule/hohlraum target as the low-foot NIC design. The key pulse-shape

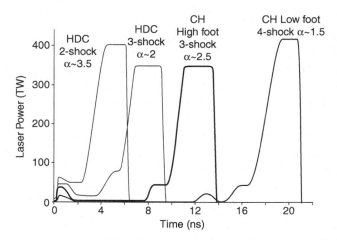

Fig. 8.   Laser pulses and resulting shock sequences and adiabats for ignition relevant capsule designs discussed in this section. The pulse shapes used for the HDC capsules are short enough that near vacuum and low density gas-filled hohlraums may be used.

changes, as compared to the low-adiabat, low-foot pulse used for the NIC, are to have more laser power at an early time (the "picket"), which causes the radiation temperature in the 'foot' of the pulse to be higher. This higher drive temperature generates a stronger first shock and puts the fuel on a higher adiabat as shown in Fig. 8 below. In addition, a consequence of the higher first shock is that only three shocks instead of four are needed to reach peak pressure without a significant further increase in fuel entropy.

The higher foot temperature and resulting higher adiabat causes more ablative stabilization, longer scale lengths and lower In-Flight-Aspect-Ratio, all of which reduce the Rayleigh–Taylor growth. The impact of RM growth is also reduced. The RM perturbations from the shock driven by the foot oscillate as they pass through the ablator. For the high-foot pulse shape, these oscillations change the sign of initial perturbations in the middle of the surface mode spectrum, resulting in reduced growth for this part of the spectrum of initial perturbations. The reduction of growth of surface perturbations for the high-foot implosions was measured in a series of hydro-growth experiments.[84–87] However, the high-foot pulse also results in a less compressed final fuel assembly reducing ideal margin for ignition in 1D.

A series of DT-layered high-foot implosions were carried out in 2013–2014. In general, these implosions performed closer to simulations in their observed yield, as well as estimated hot spot density and peak stagnation pressure. They have also been diagnosed to have very low mix of ablator into the hot spot. As of the end of 2014, the highest DT yield obtained was 9.3×10^{15} neutrons, or 26 kJ of fusion energy, a factor of 10 × increase over the previous highest performing low-foot shot. At this yield 5.2 kJ is released in alpha-particle kinetic energy. Most of the alpha particles are stopped in the hot spot and contribute to additional self-heating, which further boosts the hot spot energy and neutron yield. For this shot we estimate that the total yield was more than doubled due to the effect of self-heating.[82] These experiments are thus entering the regime where alpha-heating increasingly dominates the hydrodynamics and energy balance in the hot spot (see Fig. 10). The best of these implosions had estimated peak stagnation pressures in excess of 200 GBar, compared to the ~330 GBar needed for ignition at the NIF scale for radiation driven implosions. The inferred levels of CH ablator mix into the hot-spots were low for the high-foot implosions, confirming the impact of the measured reduction of ablation front mix of the high-foot design. As with the low-foot NIC targets, the high-foot hohlraums also required large CBET in order to achieve adequate hot-spot symmetry. Further increases in yield will likely require improving the symmetry of the compressed fuel through improvements in the hohlraum drive symmetry. Ignition will also require increased compression with the resultant increase in the capsule convergence ratio. Improved uniformity of X-ray drive has been demonstrated for the "low-foot" pulse shape using Rugby shaped hohlraums with minimal CBET (discussed below[88]), and a campaign to use these hohlraums to improve the low-mode asymmetries for high-foot experiments is under way.

3.4 *Adiabat Shaped Implosions*

Adiabat Shaped Implosions have recently been explored for CH capsules by using pulse-shapes that are intermediate between the high- and low-foot pulses.[89-94] The goal is to achieve most of the reduction of ablation front hydro-instability growth observed in the high-foot implosions by using the early higher "picket" of the high-foot pulse, while using a lower "trough-level" to obtain nearly the same low fuel adiabat, and hence higher compression, as the low-foot (see Fig. 9a).

Figure 9b compares the simulated ablation front growth factors for the low-foot 4-shock low adiabat ($\alpha \sim 1.5$), the high-foot 3-shock intermediate adiabat ($\alpha \sim 2.5$) and the adiabat-shaped 4-shock ($\alpha \sim 1.7$) implosions for CH capsules with DT cryo-layers. These theoretical predictions have been confirmed by measurements.[84-87,89-94]

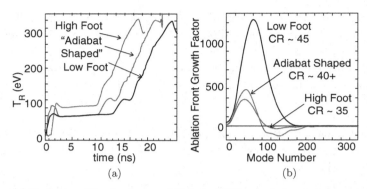

Fig. 9. (a) Radiation drive for low- and high-foot CH capsule implosions compared to adiabat shaped for a CH capsule implosion, and (b) the resulting calculated ablation front growth factor and corresponding convergence ratio (CR).

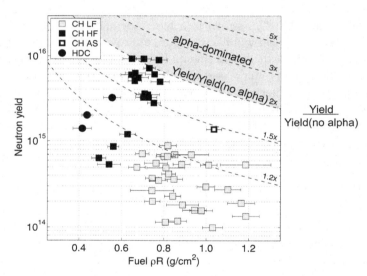

Fig. 10. Measured neutron yield vs. inferred fuel areal density for DT layered implosions on NIF. The high-foot experiments (black squares) are less compressed and have shown higher neutron yield than the low-foot NIC experiments (light grey squares). That is consistent with the high-foot higher adiabat, more stable implosion and lower convergence than low-foot. HDC (black circles) experiments fielded so far have similar lower compression, but higher velocities than high-foot leading to higher yields for similar compression. A recent adiabat shaping experiment (black square with white center) shows similar compression, but higher yield than low-foot experiments suggesting that ablation front growth was a leading cause for NIC's reduced performance.

All these designs use hohlraums that are filled with He gas at a density ranging from about 1–1.6 mg/cm^3 to keep the Au blow-off from the hohlraum walls from interfering with the laser beam propagation. The loss of energy due to absorption in the helium plasma and LPI effects such as laser backscatter instabilities from the inner beams in these hohlraums must be offset by a large amount of CBET. Although CBET provides an approach to mitigate low mode asymmetry of the final imploded state, experimental evidence indicates this has not yet been optimized to the extent required for ignition. Further improvement in implosion performance will likely require the development of alternate hohlraums configurations that reduce LPI and CBET effects.

3.5 *Rugby Hohlraum Implosions*

Changing the cylindrical shape of the hohlraum to a **Rugby-like shape** may allow uniform X-ray drive symmetry with minimal CBET. Rugby-shaped gold hohlraums driven by the NIC-style low-foot low-adiabat pulse shape have been tested on NIF. The potential benefits of a Rugby-shaped hohlraum (Fig. 11) over cylindrical hohlraum can be exercised in two ways. The reduction in surface area over a cylindrical hohlraum with similar diameter and length results in a higher coupling efficiency and achieves a near 20% improvement in peak drive X-ray flux. Alternatively, a Rugby hohlraum with a similar area to a NIC cylindrical hohlraum will achieve a similar drive but due to the larger diameter over the capsule, will have improved beam clearances (to the capsule) and higher hohlraum case-to-capsule ratio (CCR) for greater smoothing of hohlraum radiation modes and improved inner beam propagation. The Rugby target options will be further discussed below for indirect-drive targets for IFE applications.

The Rugby hohlraum experiments on NIF tested the second option.[88] With a diameter of 7 mm and a length of 10.5 mm the Rugby has essentially the same surface area (to within a few %) as the NIC cylindrical hohlraum shown in Fig 7. This results in a 22% larger CCR and nearly 30% greater beam clearance over the capsule than the NIC hohlraum. Simulations indicated, and an experiment in October 2013 demonstrated that the Rugby hohlraum can provide good inner beam propagation and X-ray drive symmetry without the large amount of CBET required to achieve adequate inner beam propagation and X-ray drive symmetry for cylindrical hohlraums. Additionally, a low X-ray flux symmetry swing in time was

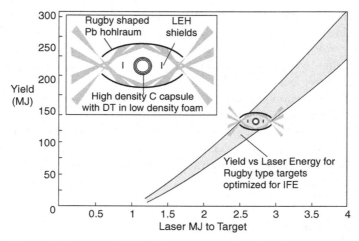

Fig. 11.   Fusion MJ vs. laser MJ for indirectly driven Rugby HSI targets optimized for IFE. The width of the band indicates uncertainty in hohlraum and capsule performance based on the assumption that current near vacuum hohlraums will perform as required for ignition. This must be demonstrated on NIF experiments.

observed, less than 1/3 of the symmetry swings for cylindrical hohlraums that employ large CBET.

3.6 *Alternate Capsule and Hohlraums*

Another way to minimize the effect of CBET, and hence improve time-dependent shape, is to reduce or eliminate the need for a hohlraum gas fill by employing shorter duration laser pulses. The He gas-fill slows down the expansion of the Au plasma blowing off the hohlraum walls.[9,50,71] This is challenging with the long pulse-lengths required for the 4-shock CH ablator design, but can be more easily achieved with the use of higher density ablators such as HDC and by reducing the number of shocks to 2 or 3. HDC, which is polycrystalline diamond grown by plasma assisted chemical vapor deposition, is an interesting option as an ablator material for indirect-drive ICF implosions. Its higher density than plastic (3.5 g/cm^3 vs. 1 g/cm^3) results in a thinner ablator for a given ablator mass with a larger inner radius for a given capsule outer radius. It exhibits higher X-ray absorption and higher overall efficiency, with a shorter laser pulse, than CH designs with the same outer radius. For this comparison, the capsule outer diameter is used because for a given hohlraum diameter, the capsule outer radius determines the degree of geometric smoothing of hohlraum radiation flux asymmetry and is usually chosen to be some maximum value. During 2013–2014, a series of experiments were carried out to examine the feasibility of using HDC as an ablator with 2, 3, and 4-shock laser pulse designs.[96]

A 2-shock high-adiabat ($\alpha \sim 3.5$) HDC capsule design was developed to reduce the ablation front growth factors by more than a factor of 10 over the ignition-scale 4-shock HDC design.[97–100] This 2-shock pulse is not an ignition design but it is predicted to produce a yield of $\sim 2 \times 10^{16}$ in 1D, which would allow valuable insight into implosion performance in the alpha-heating regime. The shorter (6–7 ns) pulse allows less time for wall motion and plasma filling and opens up the possibility of using hohlraums with intermediate gas fill (He density ~ 0.6 mg/cm^3) or near-vacuum hohlraums (NVH) with He density ~ 0.03 mg/cm^3. A density of 0.6 mg/cm^3 is about 2% of the critical density for a 0.35 μm laser while the gas fill for the standard NIF hohlraum ranges from 3–5% of the critical density. Advantages of these hohlraums, in addition to minimizing CBET, are reduced LPI effects such as reduced backscatter, little or no hot-electron generation, and hence very high (97–99%) hohlraum coupling efficiency.[97–100] Campaigns are under way to optimize hohlraum and capsule performance for 2- and 3-shock HDC designs. A 2-shock implosion in a NVH with a cryogenic DT-layered target produced 3.2×10^{15} neutrons and is a promising start (Fig. 10). Laser-to-hohlraum coupling was 98.5%, and post-shot simulations indicate yield-over-simulated performance of ~ 25–50%.[97–100] An earlier cryogenic implosion was consistent with a fuel velocity of 430 ± 50 km/sec with no observed ablator mixing into the hot spot.

3.7 *Status of ICF Implosions on NIF*

The progress of the Performance of ICF implosions can be represented in a space of yield vs. fuel ρr as shown in Fig. 10.

The yields listed include the effects of alpha particle deposition. Contours shown are lines of constant yield amplification where yield amplification is the ratio of the yield obtained including alpha deposition to the yield that would be obtained as a result of compression alone. NIF implosions using "high-foot" pulses have achieved a doubling of the yield due to alpha deposition. These implosions are now in the regime where alpha-heating increasingly dominates the hydrodynamics and energy balance in the hot spot. At ignition, an alpha particle driven burn wave, can generate yield amplifications of 100–1000. As indicated above, the best of the high-foot implosions have a peak pressure in excess of 200 GBar, a pressure that is

>60% of that needed for ignition at the NIF scale. Ongoing experiments on the NIF are working toward resolving the remaining issues on the path to ignition.

The goal for the next several years is to develop an **Ignition Relevant** hohlraum and capsule combination that minimizes the deleterious effects of laser plasma instabilities and CBET, and significantly reduces low-mode asymmetries, while maintaining high overall coupling and low hydro-instability growth. Based upon the results obtained so far, an approach being considered is a HDC capsule in a Rugby-shaped hohlraum. With the combination of shorter overall pulselength and a 3-shock adiabat 1.7–2 implosion made possible by the HDC ablator, it may be possible to substantially reduce the hohlraum gas fill (possibly to a NVH) and the adiabat 1.7–2 pulse would maintain acceptable levels of hydro-growth, although the HDC capsule has yet to be fully tested under these conditions. Coupled with the Rugby shape this should allow good inner beam coupling and X-ray drive symmetry without the need for CBET and would allow low LPI and improved overall laser target coupling. If successful, this combination is calculated to achieve ignition and net target gain on NIF.

3.8 *Extending NIF Ignition Designs to IFE*

Assuming that ignition can be demonstrated on NIF and that the target gains are consistent with those required for IFE, there are still further developments needed to have an IFE relevant laser-indirectly driven ICF target. The Au or U hohlraums used for experiments on NIF are not suitable either from the perspective of cost or nuclear radiation hazards due to activation of hohlraum materials. Additionally, the technique used to form the solid DT layers require cool-down time of 10–14 hours and place severe constraints on high repetition rate ($\sim$15 Hz) target production. Using Pb for the hohlraum material solves the cost and activation issues, and wicking liquid DT into a low-density ($\sim$20 mg/cm^3) nano-porous carbon foam shell allows rapid fill and freeze options. Experiments with Rugby targets conducted on the Omega facility showed that hohlraum radiation temperatures and implosion performance using Pb was indistinguishable from those with Au.[101] Use of the low-density foam is calculated to reduce the yield by $\leq$ 10%.[102]

Using the more efficient Rugby hohlraum discussed earlier and placing small shields between the capsule and the LEH to reduce X-ray losses, could compensate for the performance reduction. Target gains that might be realized for such IFE optimized indirect-drive Rugby targets are shown in Fig. 11.

These target gains are based on the reduced gas fill hohlraums discussed above that had high coupling efficiency and low levels of LPI effects.[103] To date, achieving these lower levels of LPI effects requires a larger size hohlraum than used for the higher fill hohlraums. Consequently the gains are lower than could be achieved using smaller hohlraums limited only by the requirements of radiation flux symmetry in the absence of LPI effects.[104] Ideas have been proposed to condition laser beams to reduce LPI effects that might make it possible to use such smaller hohlraums, and achieve higher gains. However, at this time, the target gains of Fig. 11 are the gains used in Section 5 for the Laser Indirect-Drive Inertial Fusion Power Plant.

Once ignition and net gain with Au or Au/U hohlraums and CH, HDC or Be capsules have been demonstrated on NIF, variants such as the Pb based Rugby target with LEH shields and HDC ablator with liquid DT in an inner low density ($\sim$20 mg/cc) nanoporous carbon foam optimized for IFE systems could be evaluated on NIF. NIF has the potential to be modified to allow operation at up to 2.5 MJ at 0.35 μm and 3.3 MJ at 0.53 μm. Thus, in principle, targets adequate for Inertial Fusion Energy (IFE) systems could be explored on NIF if robust ignition can be achieved with current non-IFE target configurations at $\sim$1.5–2 MJ.

4 IFE Systems

4.1 *Review of IFE basics*

Once IFE-relevant indirectly driven ICF target performance is available, the extension to an IFE power plant is challenging but — conceptually — relatively straightforward.[1] Referring to Fig. 12, the indirect-drive target is injected into a fusion chamber at a repetition rate ω (Hz). The laser with drive energy E_d (also at ω) is focused on the target producing the fusion yield E_F and a fusion power of $E_F \omega$. (For simplicity the fusion chamber is shown with just one port. In an actual fusion chamber, the target would be injected into a separate port, and there would be 48 laser ports arranged in a NIF-like configuration.) The target gain G is E_F/E_d. With an indirect-drive target $\sim$20% of the fusion output comes in the form of ions and X-rays ($\sim$10% each). The fusion chamber is filled with Xe at a density of $\sim$5–6 μg/cm^3. This gas fill has negligible impact on the propagation of the laser beams for the proposed dimension of the fusion chamber (radius 5–6 m) but

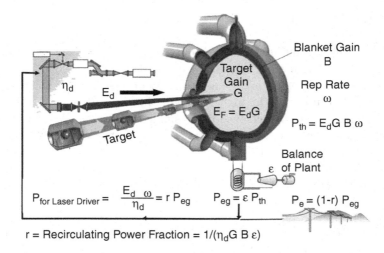

Fig. 12.   Basic parameters of an Inertial Fusion Energy power plant.

will stop 100% of the ions within a few tens of cm and approximately 90% of the X-rays, but is low enough to have no impact on the cryo DT layer of the capsule, since it is protected by the thermally robust hohlraum and thin IR reflecting membranes over the LEHs (this is discussed further in Fusion Target Systems in Section 5). The remaining X-rays are absorbed in the first wall. (For the remainder of Sections 4 and 5, we will use "Chamber" to mean "Fusion Chamber")."

After passing through the first wall and the first wall coolant, the remaining neutrons are stopped in a flowing liquid blanket. The coolant in the first wall as well as the blanket can be liquid lithium, lithium-bearing liquid metals, or lithium-bearing molten salt. The neutron–lithium interactions produce the T needed for the DT targets. There is also a net chamber energy gain (often called the Blanket Gain, B). It is defined as the ratio of the sum of the nuclear heating (neutrons and neutron-induced gamma-rays), X-ray heating, and ion heating to the initial energy of 17.6 MeV that is released from every fusion reaction. The thermal power generated is therefore $P_{th} = E_d GB\omega$.

By running the first wall coolant and the blanket through a heat exchanger and using the heat to drive a generator with an overall thermal to electric conversion efficiency ε, the gross electric power generated is εP_{th}. A fraction of the electric power generated is required to run the driver. If the overall "wall-plug efficiency" of the driver is η_d, the recirculating power fraction r is

$$r = (E_d\omega/\eta_d)/(\varepsilon P_{th}) = (E_d\omega/\eta_d)/(\varepsilon E_d GB\omega) = 1/(\eta_d GB\varepsilon). \qquad (7)$$

$G_{eff} = \eta_d GB\varepsilon$ is the effective system gain, and the net electric power available

$$P_e = (1 - r)\varepsilon P_{th} = (1 - r)E_F B\varepsilon\omega. \qquad (8)$$

(The small fraction ($\sim$2%) of gross electric power required to run auxiliary "Balance of Plant" equipment was — for simplicity — not included above.)

4.2 *IFE Metrics*

Common wisdom has been that target gain G > 100 is necessary for IFE, and that $\eta_d G$ is the relevant metric for judging IFE systems. Further, it often stated that $\eta_d G$ must be >10 or $\eta_d GB\varepsilon = G_{eff}$ must be >5 (since for optimally designed systems Bε is $\sim$0.5).[105,106]

Figure 13 shows (1 − r), or the fraction of electric power that can be sold, as a function of $\eta_d GB\varepsilon$ or G_{eff}. For this example, we chose a blanket gain B = 1.25, ε = 0.45 (typical values for properly optimized IFE systems)[107−114] and assumed η_d = 0.15, typical value for a diode-pumped solid-state laser (DPSSL).[115−118]

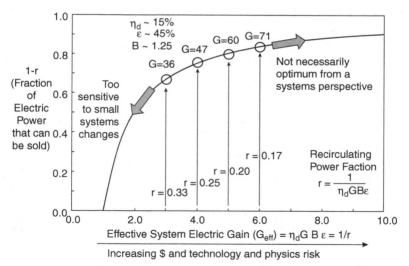

Fig. 13. A target gain G > 100 is not required, nor necessarily optimum from a systems perspective.

For $G_{eff} < 3$ (and certainly <2.5) the impact of r on the cost of electricity becomes substantial and the overall performance also too sensitive to small system changes. Similarly, once G_{eff} values of 4–5 are available, further increases are G_{eff} is not necessarily optimum from an overall systems perspective (e.g. higher risk target options, pushing the driver technology and/or operating parameters to a point where the risk of damage is increased or maintainability decreased just to gain a small increase in G_{eff}).

Clearly, for the given values of η_d, B, and ε, the higher the gain, the smaller the recirculating power fraction and the more of the gross electric power can be sold. However, we note that for indirect-drive HSI targets and DPSSL, gains >40 may be adequate and gains >60 begin to reach the point of diminishing returns. If any extra gain (and hence reduction in r) is "bought" at the expense of a larger driver (higher cost) or a target concept that has higher risk, or is incompatible with practical chamber options, a full systems optimization/evaluation is required to see if the higher G is "worth it".

We now examine the operating space for laser indirect-drive IFE systems that provide a fixed net electric power output $P_e = 1000 MW_e$.

Referring back to Fig. 12, we can show that:

$$E_F = P_e/(\varepsilon B\omega) + E_d/(\varepsilon B\eta_d). \tag{9}$$

Hence for fixed P_e and given values of ε, ω, B and η_d the relationship between the required Fusion Yield E_F and Laser Drive Energy E_d is linear.

The straight lines in Fig. 14 show the result for $P_e = 1000\,MW_e$, for $\omega = 7.5$, 10, 15 and 20 Hz, with a range of driver efficiencies for DPSSL[116–118] (from a very conservative value of $\eta_d = 0.12$, to maximum potential value of 0.18 — see Section 5 for further discussion).

We have further assumed that all systems have:

- Thermal-to-electric conversion efficiency $\varepsilon = 0.47$ (see Section 5)
- Net chamber/blanket gain B = 1.25 (see Section 5).

Also shown is the range of the E_F vs. E_d for the indirect-drive HSI Rugby targets optimized for IFE discussed in Section 3. The intersections of the target performance region with equation 10 (the "black circles" show operating points for 1000 MW_e with the specific operating repetition rate (ω) and driver (η_d). The recirculating power fraction r for the various options is indicated. We note that the systems with a driver efficiency $\eta_d = 0.15$ for DPSSLs have $G_{eff} > 4$ and recirculating power fraction r < 0.24.

A repetition rate of 20 Hz is close to a practical upper limit.[119] An examination of the constraints between target injection speed, the impact

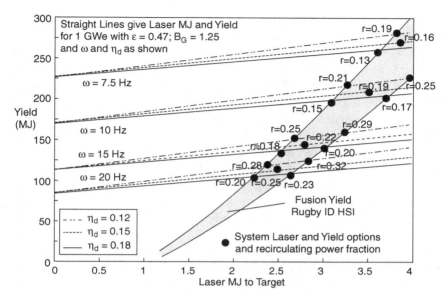

Fig. 14. Laser energy, fusion yield and recirculating power fraction for conceptual 1000-MW$_e$ IFE plants with indirectly driven Rugby HSI targets for DPSSL efficiencies (η_d) and repetition rates ω shown and expected values of **B** and ε.

of acceleration on the integrity of the DT ice layer, target "fratricide" and chamber clearing, show that it is difficult to find a self-consistent solution for repetition rates ≥ 20 Hz.

The systems with $\omega = 7.5$ Hz have lower recirculating power fractions than the systems at 15 Hz, however, the larger yields, which require larger chambers and larger laser energies, have a cost impact that negates the lower value of r.

Clearly, neither G by itself, nor η_dG nor $G_{eff} = \eta_d$ G B ε is a sufficient metric to choose between alternate IFE systems. Rather it is self-consistent target–chamber–driver physics and technology choices and combinations that matter, and systems optimizations will be required to decide on an optimum operating point.

4.3 *IFE Subsystems: Targets, Driver, Chamber, and Balance of Plant*

An IFE plant has essentially four separate "subsystems": a target system (fabrication, injection, and tracking), the driver, a chamber where the fusion reactions occur, and the "Balance of Plant" to generate electricity.

IFE Targets: In this section we will limit ourselves to discussing sub-systems relevant to indirectly driven HSI targets. For a broader discussion of IFE systems with alternative ignition targets described in Section 3 (such as direct- or indirect-drive FI, direct-drive shock ignition, laser direct drive HSI and indirectly driven HI targets) the reader is referred to Section 4 of Ref. 1 in the Suggested Reading list.

The **IFE target system** must deliver of the order of 10^6 targets per day to the center of the chamber.[119] The maximum allowable target cost will be a function of the cost of electricity. With today's prices this results in a maximum allowable cost of about 50¢ per target for a 1,000 MW$_e$ IFE plant. By using low cost materials (e.g. Pb for high-Z materials and high density C for the capsule) and adopting conventional low-cost, high-throughput, fully automated techniques several studies have estimated Rugby-like target costs to be about \$0.30/target.[120]

A combination of high-speed (200–300 m/sec), high-precision injection and tracking must be capable of determining where the target is and deliver the focused laser energy to an accuracy of a few tens of μm, or a few μradians with the anticipated stand-off distance for the final optics. The target system must also include recovery and recycling of the imploded target material residue. Significant progress has been made in recent years with the construction of prototypical gas-gun and induction injector and tracking systems at General Atomics.[121] Experiments with room temperature surrogate targets achieved injection velocities of $\sim$250 m/sec at up to 20 Hz (batch mode operation) with repeatable accuracy of delivery to chamber center of $\sim$100 μm.

The **IFE** laser **driver** must be able to deliver several MJ in a 10–20 ns precision-shaped pulse with peak powers of 400–500 TW and intensities up to 10^{15} W/cm^2 to mm-sized ICF targets.

Lasers are attractive because of their demonstrated ability to execute ICF experiments. The flash-lamp-pumped, frequency converted Nd:Glass lasers used in most ICF facilities around the world have been most intensely studied and developed. However, efficiency and repetition rate limitations make them unsuited for IFE. To overcome these limitations, research is being carried out on high average power diode-pumped solid-state lasers (DPSSL) as well as krypton-fluoride (KrF) gas lasers.

The laser medium in a **KrF laser** is a gas that can be circulated for heat removal making it possible to achieve repetition rates of 5–10 Hz. Operation at these repetition rates have been tested with the Electra KrF laser at NRL.[122–124]

KrF lasers operate at a shorter wavelength (0.25 μm) than the typical frequency-tripled wavelength (0.35 μm) of the DPSSLs, and is the preferred laser for direct-drive HSI targets (see Chapter 5). The projected efficiency of 7–7.5%[122–124] (Chapter 5), however, makes it less attractive for indirect-drive HSI.

DPSSL build on the ICF Nd:Glass laser technology (such as used for NIF), but use diodes instead of flashlamps to pump a solid-state medium, dramatically reducing the cool-down time needed between laser pulses.[115–118] Further improvement in repetition rate has been achieved using new, compact laser architectures and helium gas cooling. The Mercury laser at LLNL is a 10-Hz prototype DPSSL operating at fluence and surface heat-loads (W/cm^2) and amplifier thermal stress loads relevant for IFE. A current DPSSL laser point design for IFE applications (Section 5) has a projected efficiency of ~16–18% (electrical-to-frequency converted optical output).

Final optics present a durability issue for both KrF lasers and DPSSLs. In both systems these optics must survive the high-intensity ultraviolet laser beams and the debris, neutrons, and X-ray radiation from the fusion target. One approach uses grazing incidence metal mirrors such as aluminum-coated silicon carbide. Multilayer dielectric mirrors offer another potential option, and thin, transmissive fused silica gratings are the preferred option[88] for DPSSLs.

The basic functions for all **IFE chamber** concepts are similar.[107–114] The structure must maintain a very low density in the central cavity where the target is injected and ignited. The fusion target releases energy in the form of high-energy neutrons, X-rays, and energetic ions. Some chamber designs include mechanisms to attenuate the ions and the X-rays, but the first structural wall must (for safety and economic considerations) be designed to have a "life-time", measured in years. Various **first wall** designs have been proposed:

- Dry wall, (with or without gas fill) where the first wall is a solid structure designed to handle the fusion power. This is the simplest concept, but in the absence of a moderate density high-Z gas fill (5–10 μg/cm^3) ion spall damage to the first wall is a serious concern and may severely limit the first wall lifetime, or require very large radius chambers.
- Wetted wall, where a thin liquid layer coats the first wall and absorbs the short-range X-rays and ions before they can damage the wall, but the repeated cyclical stress of the shock-wave transmitted by the ion and X-ray-heated liquid is a serious issue.

- Thick (>50 cm lithium-bearing) liquid wall that flows between the target and first structural wall and provides protection from X-rays, ions, and neutrons. A drawback is the need for low-incidence angle driver beam illumination (half angle <20°), eliminating this chamber concept for laser indirect-drive targets, and the relatively low repetition rate (<~5 Hz) to restore the flow of the thick liquid wall between shots.

Not only must the chamber/first wall "manage" the fusion output, it must also effectively manage the intra-shot recovery — the conditions inside the chamber (such as any vapor and droplet density) that must be recovered between each shot so that the next target can be injected and the laser beams can propagate through the chamber to the target. The chamber and all of its subsystems need to be designed using materials that avoid the necessity of high-level waste disposal.

A **blanket**, at or just behind the first wall, converts the energy pulses into a steady flow of high-grade heat and breeds the T to continue to fuel the IFE plant. To accomplish these two functions, the blanket has to be thick enough to slow and absorb the neutrons to extract their energy, and it must contain lithium to react with those slowed neutrons to create the tritium. The X-rays and ions from the target would also heat the blanket (either directly or indirectly by heat transfer from any gas or liquid that attenuates/absorbs them). When lithium bearing liquids are used for tritium breeding, the liquid is generally circulated as the primary coolant for the chamber. When solid breeders such as lithium-oxide are used, high-pressure helium serves as the chamber coolant. (See Refs. 70–74 for a detailed description of tritium handling, tritium recovery and tritium breeding ratio (TBR).)

The blanket must operate at temperatures >500°C to achieve high efficiency in the **power conversion system**. With 600°C and current super critical steam technology, this provides thermal to electric conversion efficiencies of 45–47%.[107] At temperatures of 700–720°C, projected within the next 10 years an efficiency of 52–55% is possible.[112,113] More advanced technologies such as supercritical CO_2 or closed He Brayton cycles may offer efficiencies close to 60%.[112,113]

4.4 *Self-Consistent IFE Systems*

From the description of the various target, driver and chamber options, one could construct the matrix of possible IFE systems using indirect-drive laser

Table 3: Potential Systems for Laser Indirect-Drive IFE Systems.

ICF target type	Driver	Chamber
ID HSI	DPSSL	Dry Wall w/gas fill
	KrF	Wetted Wall
		Thick Liquid Wall

targets shown in Table 3 below. (For a broader discussion of self-consistent IFE systems with alternative ignition targets (such as direct- or indirect-drive FI, direct-drive shock ignition, laser direct drive HSI and indirectly driven HI targets) the reader is again referred to Section 4 of Ref. 1 in the Suggested Reading list.

Upon more detailed examination of the target–driver–chamber options, it is clear that not all combinations are physically realistic, nor are they equally attractive or practical. An IFE system must not only be composed of self-consistent subsystems, but also respond to realistic physical constraints. If we also want an IFE system that could respond to the need for low carbon energy sources for base-load power plants (GW_e power levels) within a few decades, we require technologies that can credibly be ready at the required scale within this time frame. The impacts of consequences imposed by target–driver–chamber, self-consistency constraints are examined below.

Target constraints imposed by self-consistency considerations: ICF gains adequate for IFE ($G >\sim 40$ for laser indirect-drive targets such as the Rugby target concept described in Section 3) requires that the DT fuel be a cryogenic solid shell (or liquid DT wicked into a low density nanoporous carbon foam shell) inside the capsule ablator. For a capsule that is not thermally protected the chamber pressure must be $<10^{-3}$ Torr, for the cryo DT layer to survive the transit. (With a thin IR reflective coating, pressures up to 50×10^{-3} Torr may be possible.) About 10 Torr-m of Xe is required to range out 3.5 MeV alpha particles. Even at 50×10^{-3} Torr the ions from ~ 2200 MW of fusion (required for 1000 MW_e) would result in an erosion (spall) of ~ 1 cm/year for the first wall for a 15-m-radius chamber, and a chamber radius >100 m would be required to sufficiently attenuate the alphas and maintain the integrity of the ice layer. Although physically possible, this is not a practical option.

With a density of about 5 $\mu g/cm^3$ of Xe (or about 20 Torr for typical chamber conditions) the alphas would range out in <1 m, but the capsule would now need to be protected.[119] The Pb Rugby hohlraum used for the **indirectly driven HSI targets** provide such protection. As mentioned earlier, this gas fill has negligible impact on the propagation of the laser beams for the 5–6 m radius chamber.

We now consider **driver constraints** for indirectly driven targets. DPSSLs are obviously acceptable being an extension of the Nd:Glass laser technology used for the majority of ICF experiments over the past 40 years. With diodes replacing the flashlamps and He gas cooling, DPSSLs are potentially capable of operating at 10–20 Hz with adequate brightness (beam quality), and projected "wall-plug" efficiencies (allowing for the cooling) of ∼16% are more than adequate for IFE given the gain curves shown in Fig. 14.

Even though the lower efficiency of KrF would make it less attractive for IFE than DPSSL systems with indirect-drive HSI targets, the intrinsic limit of ∼7.5% would not in and of itself rule out this combination. However, the relatively low saturation intensity of KrF lasers limits the intensity out of the KrF amplifiers to about 10 MWcm^2 at pulselengths of interest to ICF (this is to be compared with output intensities in the multi GW/cm^2 range for DPSSLs). Higher intensities are possible for KrF, but only at the cost of reduced overall efficiency. Intensities ∼10^{15} W/cm^2 are required both to focus the laser beams through the LEH of the indirect-drive hohlraum and to obtain the required radiation temperatures. Additionally for IFE, these intensities must be provided by final focusing optics that are >20 m away to survive target neutrons and X rays. The combination of lower efficiency and the constraints imposed by the intrinsic brightness of KrF lasers make them (in our judgment) less attractive than DPSSLs for indirect-drive HSI targets for IFE.

Chamber constraints. Thick (>∼50 cm) liquid wall chambers would allow actual structural first wall to survive a 50 years plant-life time. A drawback, however, is the requirement of low-incidence angle driver beam illumination (half angles ideally less than ∼20°). This eliminates this chamber concept for Rugby-style laser indirect-drive targets.

Wetted wall chambers, where a thin liquid layer coats the first wall and absorbs the short-range X-rays and ions before they can damage the wall, are an interesting concept, but the repeated cyclical stress of the shock-wave transmitted by the ion and X-ray-heated liquid is a potential issue for first wall lifetime.

With these constraints and limitations, and a goal of IFE systems that could respond to low carbon options for base-load power plants by mid-21st

Table 4: Not All Combinations are Physically Realistic or Self Consistent.

ICF target type	Driver	Chamber
ID HSI	DPSSL	Dry Wall w/gas fill
	~~KrF~~	~~Wetted Wall~~
		~~Thick Liquid Wall~~

century the possible combinations of Table 3 are reduced to DPSSL and Dry Wall w/gas fill as shown in Table 4:

In summary, we judge that, given today's knowledge and the assumption of success of the ongoing NIF experiments, indirect-drive HSI targets (Fig. 11) driven by a $\sim$2.5–3.3 MJ DPSSL operating at $\sim$15 Hz and a dry wall chamber filled with about 5–6 μg/cm^3 of Xe, would have adequate performance for Inertial Fusion Energy. Section 5 describes such an IFE system.

5 Progress towards Technologies for Laser Indirect-Drive IFE

A fusion power plant must meet a number of top-level requirements consistent with commercial operation. These include standardized, proven technology, maintainability, and constructability, a high level of quality assurance, competitive economics, and environmental sustainability. The complex set of interrelated performance requirements presents major challenges to demonstrating IFE as a commercially attractive energy source. Fortunately, significant advantage can be taken from the inherent "separability" of the IFE subsystems and plant components. Once ignition and gain has been achieved, the IFE system described below would build on demonstrated ICF physics and credible extensions of current driver and materials technologies and the science and technology for an integrated demonstration can be developed at the modular level in appropriately scaled facilities as discussed below.

The laser indirect-drive IFE power plant concept discussed below (Fig. 15) comprises a 16 Hz, DPSSL with a net "wall-plug" efficiency of approximately 15% (accounting for cooling), a target factory, a target chamber surrounded by a lithium blanket to convert the fusion power to thermal power and also breed the T needed, and the balance of the plant (heat exchange and thermal to electric conversion systems).[107−114] The system is designed for Rugby HSI targets indirectly driven by 2.75 MJ of 0.35 μm laser energy to produce a gain of about 50 and fusion yields of about 135

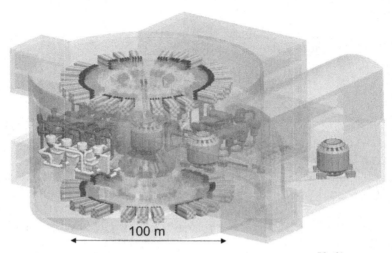

Fig. 15. A conceptual 1000-MW$_e$ laser indirect-drive IFE plant.[56-61] The vacuum chamber is in the center of the drawing. The two other vacuum chambers show how they would be removed for rapid replacement of the first wall. The laser bays are the circular areas above and below the chamber. The heat exchange systems are to the left. The target "factory" is behind the chamber.

MJ (see Fig. 11) to provide 2,165 MW of fusion power. With a blanket gain of 1.25 and a supercritical steam cycle thermal to electric conversion efficiency of $\sim$47%, this laser indirect-drive IFE plant delivers a net electric output of 1,000 MW$_e$ with a recirculating power fraction of 0.21 or an effective gain $G_{\text{eff}} \sim 5$.

Laser indirect-drive target. The indirect-drive HSI targets and fusion yield as a function of laser energy were described in Sections 2 and 3, as well as the Rugby Pb-hohlraum target that will be used for this laser indirect-drive IFE system, and will not be further discussed.

Chamber and thermal-to-electric conversion system. Figure 16 shows a model of the vacuum vessel with the first wall, blanket, and support structure (these combine to form "the chamber") sitting inside.[107] The chamber consists of eight identical sections, which would be factory built and shipped to the power plant site. Liquid lithium is the primary coolant for both the first wall and blanket.[107-113] The two systems are independently plumbed to allow greater flexibility in optimizing flow rates and coolant temperatures. Lithium is a low-activation coolant that offers excellent tritium breeding capability.

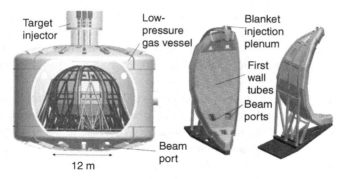

Fig. 16. The chamber has eight identical modules assembled into $\frac{1}{4}$-sections for transport to the target bay. Modules are made of steel tubes mounted to coolant plena on the sides of the blanket.

The chamber structural "first wall" would be made of 10-cm-diameter, 1-cm-thick tubes of ferritic martensitic steel such as 12YWT or oxide-dispersion strengthened ferritic steel (ODS-FS)[107] (Fig. 16). A pipe-based first wall was chosen, in part because of the high strength-to-weight ratio and ease of fabrication. The void swelling lifetime of ferritic-martensitic steels is likely to be more than 100 dpa or >four full power years for a 1000-MW$_e$ system, at which point the 1st wall sections would be replaced. The modular construction and assembly makes it possible in principle, to replace first wall/chamber sections in less than a month, and treat the replacement of first wall/chamber sections as a maintenance process. The deciding metrics are that the cost of the replacement and impact on the plant availability be economically acceptable. The first wall/chamber concept lends itself to this "exchange philosophy" and has a negligible impact on the cost of electricity. The chamber and beam path will be filled with xenon at about 6 μg/cm^3 to absorb $\sim$90% of the X-ray energy and range out the ions emitted by the target. The hot gas will then cool via radiation over 100s of μsec, sufficiently long to prevent damage to the first wall tubes. For a 135-MJ, 16 Hz, Rugby target yield and a 6-m-radius chamber, the first wall tubes will only experience a 250°C temperature spike. The gas will be pumped out between shots through the laser beam entrance ports. A chamber clearing capability that removes just 1% of the chamber contents per shot is sufficient to remove target debris for disposal or possible recycling.

This chamber design produces sufficient tritium without the use of beryllium or lithium isotopic enrichment. The tritium breeding ratio (TBR) the blanket gain B can be traded off against each other. A TBR of 1.59 could be achieved with a B of 1.10. However, this excess TBR can be traded for additional blanket gain, and the point design B (including penetrations for beamports, target injection and pumping), of 1.25 still provides adequate TBR.[107]

The chamber and energy conversion system blanket operates at a temperature of 600°C, and with current super-critical steam technology results in a thermal-to-electric conversion efficiency of 47%.[107] Second generation systems with more advanced materials (such as ODS-FS) could operate at 700–720°C and super critical steam systems are projected to achieve electric conversion efficiencies of 52–55%.[113] More advanced technologies such as supercritical CO_2 or He Brayton cycles offer the possibility of thermal efficiencies close to 60%.[113]

Laser systems for a 1,000 MW$_e$ plant. In some respects, lasers such as NIF and LMJ are prototypes for an IFE laser. Table 5 compares key laser parameters.

The laser consists of a large number of independent beamlines, each using a multi-pass architecture to optimize performance. This choice would allow reuse of much of the NIF technology and manufacturing base. The laser design differs in several respects, however, high average power operation is enabled by replacing passive cooling system with high-speed helium gas to remove heat from active components. To achieve a net "wall-plug" laser efficiency from electrical power to the frequency converted third harmonic laser light at 0.35 μm including cooling of ~15%, the flashlamps

Table 5:   Laser Parameters for NIF and a Laser for an IFE System.

Parameter	NIF	IFE Laser
Laser Pulse Energy/Peak Laser Power	1.8 MJ/500TW	2.2-3.3 MJ/500TW*
# of Beamlines /# Laser Ports	192/48	384–576/48
Port-to-port energy variation	<4% rms	<4% rms
Beam pointing error	<50 microns rms	<100 microns rms
Repetition Rate	>3 shots per day	16 Hz
Electrical-to-3ωOptical Efficiency (w/o cooling)	—	>18%
Laser-System Availability	>98.9 % of shots	>99% of time
Lifetime	>3 × 10^4 shots	>3.0 × 10^{10} shots

*Depending on the ultimate target performance of Pb-Rugby HSI targets. Systems studies show that if target gain 50 requires 20% higher laser energies, i.e. 3.3 MJ rather than the 2.75 MJ assumed in Fig. 11, the cost of electricity would increase by less than 10%.

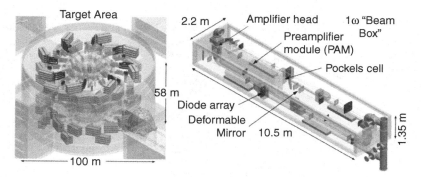

Fig. 17. The "1ω beam boxes" and fusion chamber building.

are replaced by laser-diodes. An increase in repetition rate by nearly five orders of magnitude results in average output power of order 100 kW per beamline. Many of the other technologies required for high-average-power DPSSLs, such as the thermal management of the optics, use of adaptive optics to correct thermal wavefront distortions, and methods for harmonically converting high-average-power laser, have been demonstrated with LLNL's Mercury laser.[117]

The main building block of the laser is the "1ω Beam Box",[118] which generates 8 kJ of 1.05-μm laser energy with a "wall-plug" efficiency of 25% (see Fig. 17). The 1ω beam box is the largest line-replaceable unit (LRU) within the laser system, however, it is small enough to be transported from factories to the power plant with trailers and to be handled gracefully by installation and removal equipment in the laser bay. Figure 17 shows how, for a 2.2 MJ system, 384 of these boxes are arrayed in a compact laser bay, and stacked in 4-high × 2-wide arrays arranged in circles centered on the chamber. For a system that could deliver up to 3.3 MJ, 574 boxes would be stacked in 4-high × 3-wide arrays. The 1ω beam boxes and other LRUs can be inserted and removed without disturbing neighboring beamlines. Using this approach, the laser system can achieve high system availability (>99%) with moderate beamline lifetime (MTBF >2000 hours.) and replacement time (4 hours).[118]

Final transport optics. The final transport optics represents a fundamental change relative to current ICF lasers. A set of pinholes in thick lead sheets transport the laser beam while filtering neutrons from the chamber to levels acceptable for workers (~0.04 rem/year).[111] The final optic, which is directly exposed to the output of the target, experiences the greatest

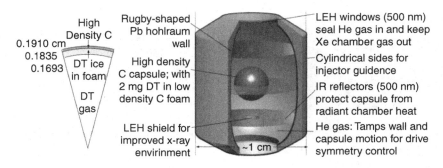

Fig. 18.   Details of the Rugby indirect-drive target.

threats. The ions and X-rays are absorbed by the xenon in the chamber and the beam tubes, so the main challenge is the 14-MeV neutrons with an average exposure of 1.5×10^{17} n/m^2 sec. The final optic must efficiently transmit the 0.35-μm laser light and allow high reliability operation, rapid replacement, and adequate mean-time-between-failure. The architecture that meets these requirements is a thin focusing grating (5-mm fused silica) that focuses and deflects the beam to target.[125] Irradiation studies of fused silica indicate that the neutron-induced absorption saturates to acceptable levels and allows efficient transmission of the beam to the target. The relatively small size and weight of the optic enable its automated extraction and replacement.

Fusion target systems. The primary challenge is to develop low cost fabrication techniques for the high-precision target shown in Fig. 18. The manufacturing strategy is to use conventional low-cost, high-throughput, fully automated techniques and to increase the batch size of chemical processes to reduce cost and increase throughput. An example is the use of die-cast hohlraum parts and an ultra-large batch CVD diamond coating process for the capsule ablator and membranes. The choice of materials involve consideration of several factors: the implosion physics; material cost and availability; potential for developing a low cost fabrication technique; thermal, mechanical and optical properties to withstand the forces of manufacturing and injection into the hot chamber; ability to recover the post-explosion debris and tritium; limiting laser propagation interference and enable recycling or disposal of waste products. By limiting the number of different low-cost materials used, the set of processes that must be developed can be reduced. The bulk material set can be reduced to lead and carbon-based (CVD diamond, and graphene-oxide, DCPD) materials.

DT is added to the capsule and small amounts of metals for dopants and reflectors.

Costs for the targets of Fig. 18 were estimated for each of the manufacturing processes. The numbers of machines for each process along with the floor space needed, the work for each process and the associated capital, consumables and personnel cost have been estimated in consultation with industry. The contribution of each process to the total target cost was evaluated for a repetition rate (PRF) of 15 Hz, resulting in a total target cost of 23¢.[120] The total contribution of the materials to the overall target cost is small ~13%.[120] The development of 3D printing manufacturing techniques would likely have a positive impact on the actual future costs.

The integrated process of **target injection, in-flight tracking and beam engagement** is a key component.[120,121] The injector must safely accelerate the target to 250–300 m/sec and deliver it to the chamber center to ±500 μm with a tilt <40 mrad in order for the laser to "hit" the target to the required ±100 μm. The target tracking and engagement system must track the position of the target to ±50 μm to allow the laser to point to the target to this ±100 μm requirement. Electromagnetic steering would likely to be used to improve the accuracy of the target trajectory. An optical tracking system would measure trajectory, velocity and tilt. The output of each laser beam to be focused on the target would be stabilized with a high-bandwidth pointing loop. The loop would be closed on an alignment beam co-aligned with the 16-Hz pulsed laser target beam. Rotating shutters would be used to shield the injector and steering equipment and especially the targets and the DT ice layer from radiation damage and heating from the fusion output from the previous target. Ongoing improvements in sensors and data processing (Speed and accuracy) will need to be exploited for IFE development.

6 Conclusion

After nearly 40 years of R&D, IFE could be at a key juncture. Demonstration of fusion ignition and gain on NIF would provide the motivation and basis for the transition from research focused on fusion to developing and demonstrating the science and technology required for an attractive commercial system.

The "separability" and modularity of the sub-systems an IFE power plant, means that the science and technology for an integrated demonstration could be developed at the modular level in appropriately scaled facilities. Optimizing and understanding target performance and its robustness at full yield and laser energy may be done independently on NIF on a single shot basis with required improvements in laser performance. Demonstration of mass production techniques for the targets at required precision and cost scalability can be done "off-line". Target delivery, tracking and target engagement, as well as chamber clearing could be demonstrated with surrogate targets and low power lasers in separate facilities. The DPSSL technology would be demonstrated with a single 100-kW beamline.

Finally, a sub-scale net 1–5 MWe engineering demonstration system using a $^1/_2$-scale laser operating at 0.1–5 Hz would allow operation and testing of 1st wall components, liquid lithium coolant and energy conversion systems in a $^1/_2$–$^1/_3$-scale chamber (see Fig. 19).

Thus once ignition and net gain is demonstrated, independent technology demonstrations would allow subscale integrated performance testing of indirect-drive sub-systems before committing to a commercial indirect-drive IFE plant.

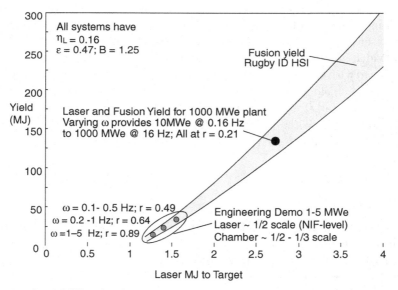

Fig. 19.   A 1–5 MWe subscale engineering system could be operated at 0.1–5 Hz with a $\sim\!^1/_2$-scale laser and less than $^1/_2$-scale fusion chamber. The full system could be brought from 10 MWe to full 1,000 MWe operation by gradually increasing the pulse repetition rate to the nominal 16 Hz.

References and Suggested Reading

Fusion as Part of a Global Energy Strategy

References

1. E. Storm (2013). *Progress Towards Inertial Fusion Energy*, Chapter 8, The World Scientific Handbook of Energy, World Scientific Publishing Company.
2. T. Anklam *et al.* (2011). The case for early commercialization of fusion energy. *Fusion Sci. Technol.*, **60**, 66–71.
3. R. J. Hawrykuk, S. Batha, W. Blanchard *et al.* (1998). Fusion plasma experiments on TFTR: A 20 year retrospective. *Phys. Plasmas,* **5**, 1577–1589.
4. A. Gibson, and the JET Team (1998). Deuterium-tritium plasmas in the Joint European Torus (JET): Behavior and implications. *Phys. Plasmas,* **5**, 1839–1847.
5. R. J. Goldston and M. C. Zarnstorff (2013). *Magnetic Fusion Energy*, Chapter 7. The World Scientific Handbook of Energy, World Scientific Publishing Company.

Physics of Ignition and High Energy Gain

ICF and Indirect-drive Target Physics

6. E. Storm, J. D. Lindl *et al.*, Progress in Laboratory High Gain ICF and Prospects for the Future. UCRL-99427, OSTI ID 6664732, NTIS No. DE89002664 (August 1988).
7. J. D. Lindl (1998). *Inertial Confinement Fusion*AIP Press:Springer Verlag, (Has a wealth of references for all aspects of ICF target physics.)
8. J. D Lindl *et al.* (2014). Review of the National Ignition Campaign 2009–2012. *Phys. Plasmas,* 2014, **21**, 020501; doi.org/10.103/1.4865400; J. D Lindl *et al.*, Erratum: Review of the National Ignition Campaign 2009–2012. *Phys. Plasmas,* **21**, 129902.
9. J. D. Lindl (1995). Development of the indirect drive approach to inertial confinement fusion and the target physics basis for ignition and gain. *Phys. Plasmas,* **2**, 3933.

Direct Drive Target Physics

10. C. D. Zhou and R. Betti (2007). Hydrodynamic relations for direct-drive fast-ignition and conventional inertial confinement fusion implosions. *Phys. Plasmas,* **14**, 072703.
11. V. N. Goncharov, T. C. Sangster, R. Betti *et al.* (2014). Improving the hot-pressure and demonstrating ignition hydrodynamic equivalence in cryogenic deuterium-tritium implosions on Omega. *Phys. Plasmas,* **21**, 056315.
12. T.C. Sangster, V. N. Goncharov, R. Betti *et al.* (2010). Shock-tuned Cryogenic-deuterium-tritium Implosion Performance on Omega. *Phys. Plasmas,* **17**, 056312.

13. R.L. McCrory, R. Betti *et al.* (2013). Progress towards polar-drive ignition for the NIF. *Nucl. Fusion*, **53**, 113021.
14. T. J. B. Collins, J. A. Marozas, K. S. Anderson *et al.* (2012). A polar-drive-ignition design for the National Ignition Facility. *Phys. Plasmas*, **19**, 056308.
15. P. W. McKenty, V. N. Goncharov *et al.* (2001). Analysis of direct-drive ignition capsule designed for NIF. *Phys. Plasmas*, **8**, 2315.
16. Y. Kato, K. Mima *et al.* (1984). Random phasing of high-power lasers for uniform target acceleration and plasma–instability suppression. *Phys. Rev. Lett.*, **53**, 1057.
17. R. H. Lehmburg, A. J. Schmitt and S. E. Bodner (1987). Theory of induced spatial incoherence. *J.Appl Phys.* **62**, 2680.
18. S. Skupsky *et al.* (1989). Improved laser-beam uniformity using the angular dispersion of frequency-modulated light. *J. Appl. Phys*, **66**, 3456.
19. S. E. Bodner *et al.* (1993). Krypton-fluoride laser fusion development in the USA, Plasma Physics and Controlled Nuclear Fusion Research 1992 (International Atomic Energy Agency, Vienna), **3**, 51.
20. S. E. Bodner *et al.* (2002). Overview of new high gain target designs for a laser fusion power plant. *Fus. Engin. Design*, **60**, 93.
21. D. G. Colombant *et al.* (2007). Direct-drive laser target designs for sub-megajoule energies. *Phys. Plasmas*, **14,** 056317.

Heavy Ion and Iodine Gas Laser ICF

22. S. Yu, W. Meier, R. Abbott *et al.* (2003). An updated point design for heavy ion fusion. *Fusion Sci. Technol.*, **44**, 266–273.
23. B. Logan, L. Perkins and J. Barnard (2008). Direct drive heavy-ion-beam inertial fusion at high coupling efficiency. *Phys. Plasmas*, **15**, 072701.
24. G. G. Kochemasov (1994). Laser fusion investigations on high power iodine laser ISKRA in Arzamas-16. *Proceedings of the 23rd ECLIM*, St. John's College, Oxford England.

Fast Ignition ICF

25. M. Tabak, J. Hammer, M. Glinsky *et al.* (1994). Ignition and high gain with ultrapowerful lasers. *Phys. Plasmas*, **1**, 1626–1634.
26. N. Basov, S. Yu and L. Feokistov (1992). Thermonuclear gain of ICF targets with direct heating of ignitor. *J. Sov. Laser Res.*, **13**, 396–399.
27. R. Kodama, P. A. Noreys, K. Mima *et al.* (2001). Fast heating of ultrahigh-density plasma as a step towards laser fusion ignition. *Nature*, **412**, 798.
28. S. Atzeni, A. Schiavi and C. Bellei (2007). Targets for direct-drive fast ignition at total laser energy of 200–400 kJ. *Phys. Plasmas*, **14**, 052702.
29. R. Betti *et al.* (2006). Gain curves for direct-drive fast ignition at densities around 300g/cc. *Phys. Plasmas*, **13**, 100703.
30. M. H. Key (2007). Status and prospects for fast ignition inertial fusion concept. *Phys Plasmas*, **14**, 055502.

31. Fusion Science and Technology. Special Issue on Fast Ignition, Volume 49, No. 3.

Shock Ignition ICF

32. R. Betti, C. D. Zhou, K. S. Anderson *et al.* (2007). Shock ignition of thermonuclear fuel with high areal density. *Phys. Rev. Lett*, **98**, 155001.
33. L. J. Perkins, R. Betti, W. H. LaFortune *et al.* (2009). Shock ignition: A new approach to high gain inertial confinement fusion on NIF. *Phys. Rev. Letts*, **103**, 045004.
34. A. J. Schmitt *et al.* (2010). Shock ignition target designs for inertial fusion energy. *Phys. Plasmas*, **17**, 042701.
35. A. J. Schmitt *et al.* (2011). Status of directly-driven shock ignition target designs. *Bull. Am. Phys. Soc.*, **56**, 221.
36. S. Atzeni, X. Ribeyre, G. Schurtz *et al.* (2014). Shock ignition of thermonuclear fuel: Principles and modeling. *Nucl. Fusion*, **54**, No. 5, 054008.
37. K. S. Andersen, R. Betti *et al.* (2013). A polar-drive shock-ignition design for the National Ignition Facility. *Phys. Plasmas* **20**, 056312.

World-wide ICF Laser Facilities and ICF Research

38. K. A. Tanaka, T. Yamanaka *et al.* (1995). Cryogenic deuterium target experiments with the GEKKO XII green laser system. *Phys. Plasmas*, **2**, 2495.
39. K. Mima, K. Kato *et al.* (1996). Recent progress of implosion experiments with uniformity-improved *GEKKO XII* laser facility at the Institute of Laser Engineering, Osaka University. *Phys. Plasmas*, **3**, 2077.
40. K. Mima *et al.* (2004). Present status and future prospects of laser fusion and related high energy density plasma research. *AIP Conf. Proc.*, **740**, 387.
41. C. B. Edwards, C. N. Danson *et al.* (1998). 200 TW upgrade of the Vulcan Nd:glass laser facility. *AIP Conf. Proc.*, **426**, 485.
42. N. C. Woolsey *et al.* (2001). Collisionless shock and supernova remnant simulation on Vulcan. *Phys. Plasmas*, **8**, 2439.
43. M. Koenig, A. Benuzzi-Mounaix *et al.* (2006). High energy density physics on LULI2000 laser facility. *AIP Conf. Proc.*, **845**, 1421.
44. S. H. Batha *et al.* (2008). TRIDENT high-energy-density facility experimental capabilities and diagnostics. *Rev. Sci. Instr.* **79**, 10F305.
45. K. N. Moncur *et al.* (1995). Trident: A versatile high-power Nd:glass laser facility for inertial confinement fusion experiments. *Applied Optics*, **34**, 21.

The National Ignition Facility

46. J. A. Paisner *et al.* (1998). Conference Proceedings. LEOS'98. 11th Annual Meeting, IEEE Lasers and Electro-Optics Society 1998 Annual Meeting (Cat. No.98CH36243): vol. 1, 390–391.
47. G. H. Miller, E. I. Moses and C. R. Wuest (2004). The national ignition facility: Enabling fusion ignition for the 21st century. *Nucl. Fusion*, **44**, S228.

48. E. M. Campbell, N. C. Holmes, S. B. Libby *et al.* (1997). The evolution of high-energy-density physics: From nuclear testing to the superlasers. *Laser and Particle Beams*, **15**(4), 607–626.

49. Committee for a Second Review of DOE's ICF Program, Final Report — Second Review of the Department of Energy's Inertial Confinement Fusion Program, National Academy of Sciences, Commission on Physical Sciences, Mathematics, and Applications, National Research Council, National Academy Press, Washington, D.C. (September 1990)

50. J. D. Lindl *et al.* (2004). The physics basis for ignition using indirect drive targets on the NIF. *Phys. Plasmas*, **11**, 339.

51. M. Cray *et al.* (1996). Path to ignition: US indirect target physics. *AIP Conference Proceedings* (369, pt.1), 53–60.

52. B. M. Van Wonterghem *et al.* (1995). *Proc. SPIE 2633, Solid State Lasers for Application to Inertial Confinement Fusion (ICF)*, 22 (December 8, 1995).

53. B. M. Van Wonterghem *et al.* (1997). *Appl Optics*, **30** (2012) 4932.

ICF Facilities under Construction as of 2015

54. B. Canaud *et al.* (2002). Laser Mégajoule irradiation uniformity for direct drive. *Phys. Plasmas*, **9**, 4252; http://dx.doi.org/10.1063/1.1504102

55. C. Lion *et al.* (2010). The LMJ program: An overview. *J. Phys.: Conf. Ser.*, **244 Part 1**, 012003.

56. C. Cavallier *et al.* (2005). Inertial fusion with the LMJ. *Plasma Phys. Control. Fusion*, **47**, B389.

57. J. Tassart *et al.* (2004). Overview of inertial fusion and high-intensity laser plasma research in Europe. *Nucl. Fusion*, **44**, S134, IAEA, Vienna.

58. Y. Pu, T. Huang *et al.* (2015). Direct-drive cryogenic-target implosion experiments on SGIII prototype laser facility. *Phys. Plasmas*, **22**, 042704.

ICF Experiments and Theory

Early ICF Research

59. Chroma: G. Charatis *et al.* (1975). Experimental study of laser driven compression of Spherical glass shells. *Proc. Of the 5th IAEA Plasma Fusion Conf. in Tokyo*, Japan 1974, Published in Plasma Physics and Controlled Nuclear Fusion Research 1974, IAEA **2**, 137 Vienna.

60. Janus: National Technical Information Service Document UCRL 5002175 (D.R. Speck, D. R, E. Storm and J. F. Swain, "Janus laser system" Laser Program Annual Report-1975, Lawrence Livermore National Laboratory, Livermore, CA (pp. 64–67). Copies may be obtained from the National Technical Information Service. Springfield, VA, 22161.

61. Cyclops: E. Bliss *et al.* (1975). "Cyclops laser system" Laser Program Annual Report-1975, Lawrence Livermore National Laboratory, Livermore, CA (pp. 69–74). National Technical Information Service Document UCRL 5002175, Copies may be obtained from the National Technical Information Service. Springfield, VA, 22161.

62. Argus: W. W. Simmons, D. R. Speck and J. T. Hung (1978). *Appl. Opt.*, **17**, 999.
63. Shiva: D. R. Speck *et al.* (1981). The shiva laser-fusion facility. *IEEE J. Quantum Electronics*, **QE-9**, 1599.
64. Helios: R. B. Perkins (1978). Recent Progress in Inertial Confinement Fusion Research at the Los Alamos Scientific Laboratory. *Proc. Of the 7th IAEA Plasma Fusion Conf. in Innsbruck*, Austria 1978, Published in Plasma Physics and Controlled Nuclear Fusion, Vienna.
65. A. Bekiarian, E. Buresi, A. Coudeville *et al.* (1978). Work on laser interaction and implosion at centre d'etudes de limeil. *Proceedings of the Seventh International Conference on Plasma Physics and Controlled Nuclear Fusion*, Innsbruck Austria.
66. F. Amiranoff, R. Fabro, E. Fabre *et al.* (1979). Experimental transport studies in laser produced plasmas at 1.06 and 0.53 μm. *Phys. Rev. Letts*, **43**, 522–525.
67. G. Velarde *et al.* (2007). *Inertial Confinement Nuclear Fusion: A Historical Approach by its Pioneers*. Foxwell and Davies (UK).

Indirect-Drive Ignition Physics on NIF

68. M. J. Edwards *et al.* (2013). Progress towards ignition on the national ignition facility. *Phys Plasmas*, **20**, 070501.
69. P. Michel *et al.* (2009). Tuning the implosion symmetry of ICF targets va controlled crossed-beam energy transfer. *Phys. Rev. Lett.*, **102**, 056305.
70. P. Michel *et al.* (2010). Symmetry tuning via controlled crossed-beam energy transfer on the National Ignition Facility. *Phys Plasmas*, **17**, 050910.
71. J. D. Lindl and E. I. Moses (2011). Special topic: Plans for the National Ignition Campaign on NIF. and the associated papers by S. W Haan, *et al.*, M. J. Edwards, *et al.*, and O. L. Landen, *et al.*, *Phys Plasmas*, **18**, 050910, 051001, 051002 and 051003.
72. S. Glenzer, B. MacGowan *et al.* (2010). Symmetric inertial confinement fusion implosions at ultra-high laser energies. *Science*, **327**, 1228–1231.
73. P. Chang, R. Betti, B, Spears *et al.* (2010). Generalized measurable ignition criterion for inertial confinement fusion. *Phys. Rev. Lett.*, **104**, 135002.
74. N. Meezan, L. Atherton, D. A. Callahan *et al.* (2010). National ignition campaign *hohlraum* energetics. *Phys. Plasmas*, **17**, 056304.
75. N. B. Meezan (2013). *Hohlraum* designs for high velocity implosions on NIF. *EPJ Web of Conferences*, **59**, 02002.
76. T. Ma *et al.* (2013). Onset of hydrodynamic mix in high-velocity, highly compressed inertial confinement fusion. *Phys. Rev. Lett.*, **111**, 085004.
77. O. A. Hurricane *et al.* (2014). Fuel gain exceeding unity in an inertially confined fusion implosion. *Nature*, **506**, 343.
78. T. R. Dittrich *et al.* (2014). Design of a high-foot high-adiabat ICF capsule for the National Ignition Facility. *Phys. Rev. Lett.*, 2014, **112**, 055002.

79. H. S. Park *et al.* (2014). High-adiabat high-foot inertial confinement fusion implosion experiments on the National Ignition Facility. *Phys. Rev. Lett.*, **112 (5)**, 055001.
80. T. Ma *et al.* (2015). Thin shell, high velocity inertial confinement fusion implosions on the National Ignition Facility. *Phys. Rev. Lett.*, **114**, 145004.
81. D. A. Callahan *et al.* (2015). Higher velocity, high-foot implosion on the National Ignition Facility laser. *Phys Plasmas*, **22**, 056314.
82. O. A. Hurricane *et al.* (2015). Alpha-particle self-heating dominated inertially confined fusion plasmas. *Submitted to Nature.*
83. T. Döppner *et al.* (2015). Demonstration of high performance in layered deuterium-tritium capsule implosions in uranium *hohlraums* at the NIF. *Phys. Rev. Lett.*, **115**, 055001.
84. V. A. Smalyuk *et al.* (2014). First measurements of hydrodynamic instability growth in indirectly driven implosions at ignition-relevant conditions on the National Ignition Facility. *Phys. Rev. Lett.*, **112**, 185003.
85. K. S. Raman *et al.* (2014). An in-flight radiography platform to measure hydrodynamic instability growth in inertial confinement fusion capsules at the National Ignition Facility. Physics of Plasmas. *Phys. Plasmas*, **2014, 21**, 072710.
86. D. T. Casey *et al.* (2014). Reduced instability growth with high adiabat ("high foot") implosions at the National Ignition Facility. *Phys. Rev. E*, **90**(1), 011102(R).
87. J. L. Peterson *et al.* (2015). Validating hydrodynamic growth in National Ignition Facility implosions. *Phys. Plasmas*, **22**, 05609.
88. P. Amendt *et al.* (2014). Rugby *hohlraum* experiments on the National Ignition Facility: Comparison with high-flux modeling and the potential for gas-wall interpenetration. *Phys. Plasmas*, **21**, 112703.
89. D. S. Clark (2014). Pulse shape options for a revised plastic ablator ignition design. *Phys. Plasmas*, **21**, 112705.
90. K. L. Baker (2015). Adiabat-shaping in indirect drive inertial confinement fusion. *Phys. Plasmas*, **22**, 052702.
91. J. Milovich *et al.* (2015). Design of indirectly-driven, high-compression ICF implosions with improved hydrodynamic stability. submitted to *Phys. Plasmas.*
92. D. T. Casey *et al.* (2015). Improved performance of high-ρR indirect drive implosions at the National Ignition Facility using a 4-shock adiabat shaped drive. submitted to *Phys. Rev. Lett.*
93. J. L. Peterson *et al.* Differential ablator-fuel adiabat tuning in indirect-drive implosions. *Phys. Rev. E*, **91**, 031101(R) (2015).
94. V. A. Smalyuk *et al.* (2015). First results of radiation-driven, layered deuterium-tritium (DT) implosions with a 3-shock adiabat-shaped drive at the National Ignition Facility. submitted to *Phys. Rev. Lett.*
95. A. G. MacPhee *et al.* (2015). Stabilization of high-compression, indirect-drive Inertial Confinement Fusion implosions using a 4-shock adiabat-shaped drive. submitted to *Phys. Rev. Lett.*

96. A. J. MacKinnon (2014). High-density carbon ablator experiments on the National Ignition Facility. *Phys Plasmas*, **21**, 056318.

97. O. S. Jones *et al.* (2015). Demonstration of near-ideal hohlaum performance at low gas-fill density. submitted to *Phys. Rev. Lett.*

98. L. Berzak Hopkins *et al.* (2015). First high-convergence cryogenic implosions in a near-vacuum *hohlraum*. *Phys. Rev. Lett.*, **114**, 175001.

99. L. Berzak Hopkins *et al.* (2015). Near-vacuum *hohlraums* for driving fusion implosions with high density carbon ablators. *Phys. Plasmas*, **22**, 056318.

100. J. S. Ross *et al.* (2015). High-density carbon capsule experiments on the national ignition facility. *Phys. Rev. E*, **91**, 021101 (R).

101. J. S. Ross *et al.* (2013). Lead (Pb) *Hohlraum*: Target for inertial fusion energy. *Sci Reports*, **3**, 1453.

102. E. Olson and R. J. Leeper (2013). Alternative hot spot formation techniques using liquid deuterium-tritium layer inertial confinement fusion capsules. *Phys. Plasmas*, **20**, 092705.

103. P. Amendt *et al.* (2015). High-density carbon ablator ignition path with low-density gas-filled Rugby *hohlraum*. *Phys. of. Plasmas*, **22**, 04070.

104. P. Amendt (2014), Private communication.

IFE Power Plant Technologies

Key Requirements for an IFE Power

105. R. Linford, R Betti *et al.* (2003). A Review of the US Department of Energy's Inertial Fusion Energy Program. *J. Fusion Energy*, **22**, 93–126.

106. R. Betti, D. Hammer, G. Logan *et al.* (2009). Advancing the Science of High Energy Density Laboratory Plasmas. Fusion Energy Science Advisory Committee report United States Department of Energy.

IFE Power Plant Components and Subsystems

107. J. F. Latkowski *et al.* (2011). Chamber design for the laser inertial fusion energy engine. *Fusion Sci. Technol.*, **60**, 54–60.

108. S. Reyes *et al.* (2013). LIFE tritium processing: A sustainable solution for closing the fusion fuel cycle. *Fusion Science and Technology*, **64**(2), 187–193.

109. S. Reyes *et al.* (2001). Safety assessment for inertial fusion energy power plants: Methodology and application to the analysis of the HYLIFE-II and sombrero conceptual designs. *Journal of Fusion Energy.*, **20**, 23–44.

110. S. Reyes *et al.* (2006). Safety and environmental considerations for laser IFE chambers. *Proc of the 29th ECLIM Conf*, June 11–16.

111. S. Reyes *et al.* (2007). Status of IFE safety and environmental activities in the US. *Nuclear Fusion*, **47**(7), 422–426.

112. W. Meier, A. Raffray, S. Abdel-Khalik *et al.* (2003). IFE chamber technology — status and future challenges. *Fusion Sci. Technol.*, **44**, 27–33.

113. W. R. Meier, K. J Kramer *et al.* (2014). Fusion technology aspects of laser inertial fusion energy. *Fusion Engineering and Design*, **89**, 2489–2492.

114. S. Sharafat *et al.* (2005). Micro-engineered first wall tungsten armor for high average power laser fusion energy systems. *J. Nucl. Matls.*, **347** 217–243.

Diode Pumped Solid State Lasers (DPSSL)

115. Y. Kozaki (2000). Way to ICF reactor. *Fusion Eng. Des.*, **51–52**, 1087–1093.
116. C. Orth, S. Payne and W. Krupke (1996). A diode pumped solid state laser driver for inertial fusion energy. *Nucl. Fusion*, **36**, 75–116.
117. A. Bayramian *et al.* (2007). The mercury project: A high average power, gas-cooled laser for inertial fusion energy development. *Fusion Sci. Technol.*, **52**, 383–387.
118. A. Bayramian *et al.* (2011). Compact, efficient laser systems for laser inertial fusion energy. *Fusion Sci. Technol.*, **60**, 28–48.

IFE Target Fabrication and Target Injection Subsystems

119. R. Miles *et al.* (2014). Thermal and structural issues of target injection into a laser-driven inertial fusion energy chamber. *Fusion Sci. Technol.*, **66(2)**, 343–348.
120. R. Miles *et al.* (2009). Laser inertial fusion energy target fabrication costs. *LLNL-TR-416932*.
121. D. T. Frey *et al.* (2005). Rep-rated target injection for inertial fusion energy. *Fusion Science and Technology*, **47**(4), 1143–1146.

KrF Laser Technology

122. J. Sethian, M. Friedman and R. Lehmberg (2003). Fusion energy with lasers, direct drive targets, and dry wall chambers. *Nucl. Fusion*, **43**, 1693–1709.
123. S. Obenschain, D. Colombant *et al.* (2006). Pathway to a lower cost high repetition rate ignition facility. *Phys. Plasmas*, **13**, 056320.
124. P. M. Burns, M. Myers *et al.* (2009). Electra: An electron beam pumped KrF rep-rate laser system for IFE. *Fusion Sci. Tech.*, **56**, 349.

Final Optics

125. J. F. Latkowski *et al.* (2003). Fused silica final optics for inertial fusion energy: Radiation studies and system-level analysis. *Fus. Sci. Tech.*, **43**, 540–558.

Chapter 5

Direct-Drive Laser Fusion

John Sethian

Laser Plasma Branch, Plasma Physics Division
US Naval Research Laboratory
4555 Overlook Ave SW, 20375 Washington DC, USA
john.sethian@nrl.navy.mil

In the core of a future direct-drive laser fusion power plant, an array of high power lasers directly compress a pea sized pellet of frozen deuterium and tritium (DT) that has been injected into a reaction chamber. The pellet reaches a high enough density and temperature to undergo a thermonuclear fusion reaction. The fusion energy is absorbed by the chamber wall and blanket as heat and converted into electricity. The process repeats 5–10 times per second. This chapter discusses the history and progress in developing this concept. Direct-drive fusion targets have been designed with computer simulations, using codes benchmarked with experiments. These predict power plant class performance (gain) at laser energies around 1 MJ. Two lasers have been developed: the electron beam pumped krypton fluoride (KrF) laser and the diode pumped solid state laser (DPSSL). KrF is predicted to have higher target performance, but both have the potential to meet the fusion power plant requirements. Credible technologies have been developed, and in many cases demonstrated, for the other needed components. This chapter also gives a brief description of the heavy ions, pulsed power, and fast ignition approaches to inertial fusion energy (IFE).

1 Introduction

Thermonuclear fusion powers the sun and the stars. If harnessed on earth, a fusion power plant would emit no greenhouse gases or chemical pollution and have a readily available fuel supply. Fusion reactions occur when two light nuclei are brought together to create a heavier one. The easiest reaction to ignite is with DT, two isotopes of hydrogen. Deuterium comes from seawater. One gallon of seawater has the energy equivalent of 300 gallons of gasoline. Tritium is produced in the power plant by transmutation of lithium, an abundant element. A fusion reactor is inherently safe and can neither explode nor meltdown. Reactivity is induced in the surrounding structure, but by judicious choice of materials the resulting waste would qualify for shallow land burial and be a readily tractable issue.

Fusion requires the DT be heated to 10 kilo electron volts (keV) = 116,000,000°C, and contained long enough for a sustained reaction to be ignited. Two fundamentally different approaches are under development: in *Magnetic Fusion Energy (MFE)*, the DT is in the form of an ionized gas, or *plasma*, and is confined by large magnetic fields that isolate it from the material walls. This is discussed in Chapter 6. As discussed in Chapter 7, the scientific principal of this approach is expected to be demonstrated by the ITER device now under construction in the South of France by an international partnership. In *IFE,* intense laser beams, particle beams, X-rays, or pulsed magnetic fields are used to very rapidly compress the DT to very high density. The compressed fuel is confined by its own inertia long enough for the thermonuclear ignition and burn to take place. The two main laser-based approaches to IFE are discussed in this volume. Chapter 4 discusses the *Indirect-Drive* approach, in which the DT pellet is suspended inside a small cavity, or *hohlraum*. The lasers heat the inner wall of the hohlraum to high enough temperatures (few hundred eV) to emit X-rays. These X-rays in turn compress the pellet by the ablation mechanism discussed below. This is the approach chosen by the National Ignition facility (NIF) in Livermore California, although achieving ignition by this approach has proven to be elusive.

In the *Direct-Drive* approach discussed in this chapter, the laser energy is deposited directly on the pellet without the intervening hohlraum. The lasers are powerful enough to blow off, or *ablate*, the outer layer of the pellet. This ablated material expands outward at high velocity. The part of the pellet left behind is driven inward by conservation of momentum (the rocket effect). Implosion velocities of 300–400 km/sec can be achieved, causing the pellet to be compressed to very high density — typically 1000× the initial solid density. Kinetic energy is converted to thermal energy when the imploding shell stagnates, and heats the center to very high temperature (~10 keV). These extreme conditions are enough to cause a thermonuclear

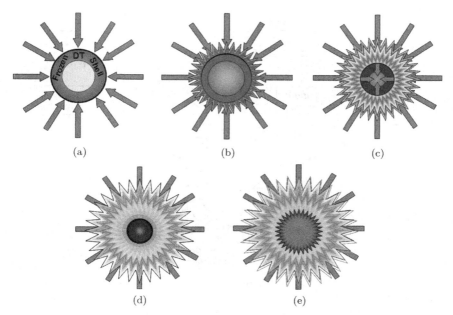

Fig. 1. Sequence of events in an energy producing inertial fusion implosion: (a) Irradiation, (b) Ablation, (c) Compression, (d) Hot spot formation, and (e) Subsequent thermonuclear burn.

fusion "hot spot" to ignite. The fusion burn then propagates though the fuel. The sequence is shown in Fig. 1.

The predicted typical fusion energy released for a direct-drive laser IFE target is in the range of 150 megajoule (MJ) to 400 MJ. In terms of a fusion power plant, a 360 MJ pellet imploded continually at 5 Hz would deliver about 685 megawatt-electric (MWe) to the grid. This assumes a thermal conversion efficiency (fusion power to electrical power) of about 40%, an energy multiplier in the blanket of 1.1 times, and a laser efficiency of 7%. The energy multiplier occurs because tritium breeding in one isotope of lithium is an exothermic reaction. The plant output can be varied by varying the laser energy, the target design, the laser efficiency, or the repetition rate. Conceptual drawings of the major components of a laser direct-drive IFE power plant are shown in Fig. 2.

No one has yet to achieve thermonuclear breakeven (the point at which the fusion energy produced by the pellet exceeds the laser energy required to heat and compress it), much less produce enough energy for a power plant or demonstrated the sequence shown in Fig. 2. Computer simulations are used

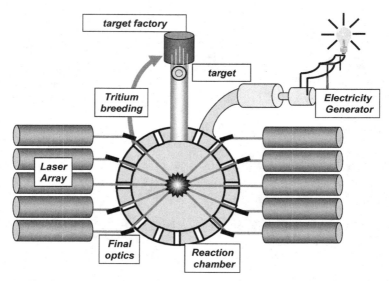

Fig. 2. The main components of a direct-drive IFE fusion chamber. The target is injected into a chamber. When the target reaches chamber center the target is symmetrically imploded by an array of laser beams. The fusion energy is collected by the chamber wall, and converted into heat by a surrounding blanket that contains lithium. The neutrons from the fusion reaction react with the lithium to produce tritium and additional energy. The process is repeated continually at repetition rates between 5 and 15 pulses/sec (Hz).

to produce "*Target Designs*" that are predicted to meet the performance requirements (fusion output) needed for a practical power plant. These become the basis for conceptual designs of a fusion power plant. The simulations use sophisticated *target design codes* that have been bench tested against experiments. Confidence is gained in the target designs by verifying that the underlying codes can predict a wide variety of experiments over a suitably broad range of parameters. Most of these experiments have been performed at laser energies well below that needed for a reactor. But the recently completed NIF laser, although configured for indirect-drive, has enough energy to address the key issues in direct-drive physics at full scale.

A key measure of the performance of an IFE target design is its *gain*, which is the ratio of the fusion energy released divided by the laser energy incident on the outside of the pellet (or in the case of indirect-drive, the hohlraum). High gain is achieved by a proper choice of target materials, composition, and dimensions, along with precise *pulse shaping*, i.e. varying the laser intensity during the laser pulse. Of course, target designs need to be carried out in concert with the other key components in a power plant, to ensure all the parts can work together as an integrated system.

2 History of Direct-Drive Laser IFE Power Plant Concepts

The concept of laser direct-drive was first published in Scientific American in June 1971, in a paper by Moshe Lubin of the University of Rochester, and Arthur Frass of Oak Ridge National Laboratory. The title was simply "Fusion by Laser". The authors stated: "Experiments indicate that energy-releasing fusion reactions can be initiated and to some extent controlled without a confining magnetic field by focusing a powerful laser pulse on a frozen pellet of fuel". The leading edge of the laser pulse vaporized the pellet, whereas the main part of the pulse heated the resulting plasma to fusion conditions. The required duration of the laser pulse was estimated to be about 0.1 nanoseconds (nsec). The reaction chamber was estimated to be 3–5 m in diameter, with the inner wall of the chamber protected from the neutrons by a vortex of liquid lithium. Bubbles in the lithium absorbed the shock of the target. The lithium also provided a blanket for tritium breeding. The repetition rate was anticipated to be a pulse per 10 sec.

Of course many of the details were a bit off, but that was to be expected considering the state of scientific understanding at the time. Nevertheless, this paper outlined the basic approach to an IFE reactor that is still being followed today. A nanosecond pulse high energy laser, with appropriate pulse shaping to control the implosion, a frozen DT target, repetitively pulsed operation, a need to protect the reaction chamber from the emissions from the target, and a means to absorb the neutrons in a tritium breeding blanket.

As shown in Table 1, the ensuing 38 years since that Scientific American article saw the development of several direct-drive laser fusion reactor concepts. In the meantime, similar "power plant studies" were carried out for other approaches to fusion energy. The primary goals of those studies were to evaluate the viability of a particular approach, to facilitate a relative comparison between approaches, and to define technologies that needed to be developed. Generally, the studies were based on conceptual untested designs for the key components as well as unproven, or sometimes even poorly understood, IFE target physics. The High Average Power Laser (HAPL) program in 2009 took a rather different approach. It used the newly developed direct-drive high performance target designs with their firmer physics base, and concentrated on developing and demonstrating the key technologies needed for a power plant. Past reactor studies were used, but

Table 1:   Major Laser Direct-Drive Reactor Concepts.

Reactor name (year)	Electric power (MWe)	Laser energy (MJ)	Laser type	Illumination geometry	Chamber wall protection/ breeder
BLASCON 1972	35	1.0	e-beam CO_2	Single sided	Liquid Lithium vortex, with bubbles
HYLIFE 1985	1010	4.5	ND: Glass	Single sided	Liquid Lithium Jets, with gas voids
SIRIUS-T 1990	507	2.0	e-beam KrF	Spherically symmetric	Xenon gas protected graphite wall, liquid lithium breeder
Prometheus L 1992	960	4.0	Discharge KrF	Spherically symmetric	Liquid Lead protection, Li_2O pellet breeder
SOMBRERO 1992	1000	3.4	e-beam KrF	Spherically symmetric	Xenon gas protected graphite wall, Li_2O pellet breeder
SOLASE 1997	1000	1.0	CO_2	Two Sided	Neon gas protection/ graphite blanket with Li_2O pellets
HAPL 2009	750	2.5	KrF or DPSSL	Spherically symmetric	Vacuum chamber, engineered Ti wall, liquid LiPb breeder

only as a guide to show which components needed to be developed. Of particular importance was the development of two laser concepts; the electron beam pumped KrF laser and the Diode Pumped Solid State Laser (DPSSL). The program showed both have the credible potential to meet the fusion energy requirements for repetition rate, durability, and efficiency. The fundamental architecture of the DPSSL laser was later adopted for the LIFE indirect-drive approach proposed by Livermore National Laboratory. The HAPL program developed viable concepts for the other key components, including: target fabrication, target injection, final optics, and the chamber. Most of the technologies were demonstrated in subscale systems. One of the hallmarks of the HAPL program was that it was the first time the technologies and the physics were developed as an integrated system. Thus, for example, target designs were developed that not only had the capability of producing high enough performance for energy production, but also could readily be fabricated with materials that could easily be recycled, and produced an energy spectrum that could repetitively be absorbed by the reaction chamber.

3 The Justification for Developing the Laser Direct-Drive Approach

There are several reasons for pursuing the laser direct-drive approach:

(i) As shown in Fig. 2, the main components are physically separated from the reaction chamber. This separation keeps the complex components such as the laser and target factory isolated from the nuclear reaction chamber. Separation also allows the components to be developed independently before integration into the system. This not only reduces development costs, it allows economical upgrades.

(ii) The laser, the most costly component (estimated to be about 1/3 the cost of the plant) is modular. It is composed of 20–60 identical beam lines. Thus all the development can be done on one laser beam line, which when perfected, is then replicated to build the entire laser system. This lowers development costs and risks.

The above advantages apply to any laser based approach to IFE. However, direct-drive also has several advantages compared to indirect drive:

(iii) Direct-drive is more efficient, hence allowing higher gains. This is because the laser energy is directly deposited onto the target rather than through an intervening hohlraum. As shown below, target designs predict power plant class gains with relatively modest laser energies of around 1 MJ with conventional direct drive, or as low as 500 kJ with "Shock Ignition." Both are described in this chapter.

(iv) Direct-drive targets are simpler to manage. They are based on spherical shells that have been repetitively fabricated in a droplet generator, they require no preferred direction of illumination, and they have no hohlraum debris to recycle.

(v) Direct-drive has simpler target physics: the main concern is hydrody-
    namic instability, which appears to be resolved with a combination
    of advances in laser technology and target design. Other issues such
    as laser target coupling and suppression of laser plasma instabilities
    (LPI) are being addressed on subscale experiments.

(vi) Direct-drive physics is easier to study and elucidate: all fusion concepts
    have complex multidisciplinary physics.The open illumination geome-
    try of direct-drive, plus its point source nature, allows ready diagnostic
    access.

4 Reactor and Target Performance Considerations

In order to gain an understanding of the performance needed for a fusion
power plant, consider a representative power flow chart, as in Fig. 3.

We assume for this example, the laser is 7% efficient, produces 1.50
MJ per pulse, and runs at 5 Hz. As discussed in the next section, the gain
of the target is taken to be 240 times. So the fusion output is 1,800 MW.
After allowing for the blanket boost (+10%) and the electrical conversion
efficiency (40%), the output power is 792 MWe. But 107 MWe is required
to drive the laser, so 685 MWe is left for the grid. The ratio of the power
required by the laser divided by the total power is 107/792 = 13.5%. This
ratio is known as the *recirculating power*, and economic considerations suggest it be as small as possible. The flow chart in Fig. 3 can be expressed in

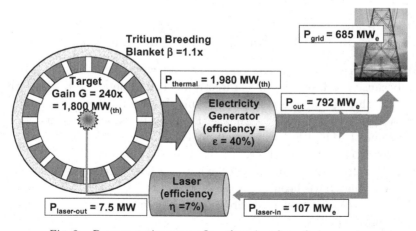

Fig. 3.   Representative power flow chart in a laser fusion reactor.

quantitative form as:

$$P_{grid} = P_{laser-in}[(G \times \beta \times \varepsilon) - 1/\eta],$$

where G is the gain, β is the blanket energy multiplication, ε is the thermal conversion efficiency, and η is the laser efficiency. Note η is the all inclusive laser efficiency: it is the laser power on target divided by the electrical power into the laser, and includes the power for cooling the laser components. The recirculating power is defined as the input power to the laser divided by the gross electrical power out, or $P_{laser-in}/P_{out} = f = 1/\varepsilon\eta\beta G$. Figure 4 shows a plot of the recirculating power fraction, f, as a function of the product of ηG, i.e. gain times laser efficiency. For this case $\beta = 1.1$ and $\varepsilon = 0.40$.

Figure 4 shows the basis of the oft-quoted requirement that ηG must be greater than 10 to have a viable inertial fusion power concept. Much less than that and too large a fraction of the power output is required to simply power the laser. However, that inequality does not tell the whole story. Gain itself is important because higher gain gives more electrical power output for a smaller (lower cost) driver. Figure 5 plots the ratio of the power delivered to the grid as a function of the power output of the laser, for various values of the laser efficiency. The power output of the laser is used because it is a measure of the size of the laser, and hence its cost. In all cases the value of $\eta G = 10$ is also plotted. Note that a 10% increase

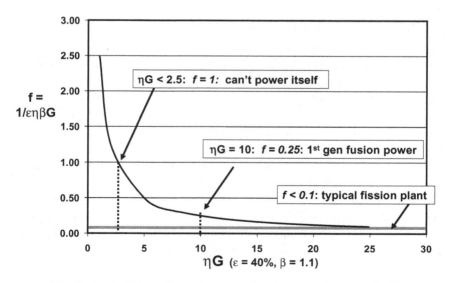

Fig. 4.   Plot of the recirculating power fraction f as a function of ηG.

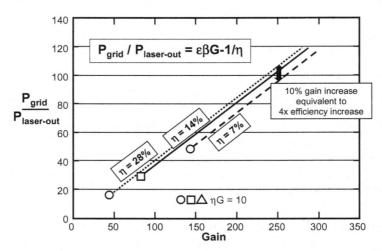

Fig. 5.   Ratio of power delivered to the grid divided by laser power, as a function of efficiency and gain.

in gain is the equivalent to a four times increase in laser efficiency. Thus, higher gain is worth more than higher laser efficiency. Put simply, gain does matter.

5 Direct-Drive IFE Class Target Designs

High gain requires that the target be compressed uniformly to high densities (typically 1,000 times the initial solid density), very rapidly (typically a few nsec), and with as small a laser as possible. The objective is to create a central hot spot that ignites the surrounding, relatively cold compressed fuel. High density is needed because the fusion reaction rate is proportional to the square of the density. Fast compression time is needed to ensure maximum density is reached before the pellet blows apart. Thermonuclear breakeven, which for IFE is defined as the point at which the fusion energy produced by the pellet exceeds the laser energy required to heat and compress it, is achieved when the product of the density n, and the confinement time, τ, in the hot spot exceeds $n\tau > 10^{14}$. This is known as the Lawson criterion. In the IFE community the Lawson criterion is usually expressed in terms of the density ρ (in g/cm^3), and fuel radius R (in cm):

$$n\tau = \rho R/(m_f c_s),$$

where m_f is the mass of the fuel and c_s is the isothermal sound velocity. The condition in which a major fraction of the alpha particles deposit their energy in the hot spot is roughly $\rho R > 0.5$.

Rapid compression is attained by maximizing the implosion velocity. The implosion velocity can be increased by increasing the drive pressure through an increase in laser intensity. But if the intensity is high enough it can exceed the threshold for LPI, which can either produce hot electrons that preheat the fuel, which decreases its density; or scatter the laser, which results in energy losses, or both. Alternatively, the implosion velocity can be increased by decreasing the target areal mass, i.e. making the target thinner but with a larger diameter. However thinner, larger diameter targets are more susceptible to hydrodynamic instabilities which can break up the fuel compression and prevent the formation of a central hot spot. The two dominant hydrodynamic instabilities are Richtmyer–Meshkov during the initial shock transit, and Rayleigh–Taylor during the acceleration. Another potential loss mechanism is cross beam energy transfer (CBET) which transfers energy from an incoming beam to an outgoing beam before it gets to the pellet. This reduces the laser absorption and coupling efficiency. Considerable progress has been made in understanding and controlling these instabilities. Of the three, hydrodynamic instabilities are the most well understood and methods have been devised and tested to minimize them. These include designing targets of specific materials, compositions, and dimensions, tailoring the laser pulse shape, and improving the laser properties such as beam smoothing. For CBET, the basic physics is understood well enough to start evaluating methods to alleviate it. These include "zooming" the laser focal spot to follow the imploding pellet and thereby reduce the beam crossing. The physics of LPI is not yet as well understood. The dependence of LPI thresholds on laser wavelength and intensity have been studied, but a predictive capability is still the subject of vigorous research.

There are three major classes of high gain laser direct-drive targets:

Conventional Symmetric Direct Drive: In these designs the illumination on target is spherically symmetric or nearly spherically symmetric. The lasers implode the target at velocities above 300 km/sec. As the imploding fuel stagnates at the center, the imploding kinetic energy is converted to thermal energy, thus creating a central spot that is hot enough and dense enough to undergo thermonuclear ignition. This launches a thermonuclear burn wave to propagate through the surrounding dense shell.

Shock Ignition: In shock ignition, which was proposed by the University of Rochester, a short duration high intensity laser pulse is applied near peak compression. This drives a high intensity shock to ignite the hot spot. Thus this separates the compression from the ignition steps. Shock ignition requires lower velocity implosions of 200–250 km/sec, i.e. only 2/3 of that required for conventional direct-drive. The illumination pattern is the same as for Conventional Symmetric Direct Drive.

Polar Direct Drive: In polar direct drive (PDD), the laser beams come in from two sides, but are adjusted in angle and intensity to give nearly spherically symmetric drive pressures. PDD will allow a variant of direct-drive to be evaluated on the NIF in its indirect-drive configuration, without going through the considerable expense of moving the beams to other ports on the NIF chamber. It also allows ready transitions between indirect-drive and direct-drive experiments. To get the required illumination symmetry, PDD intentionally increases the relative irradiation intensity at the target equator to compensate for the reduced laser coupling and reduced hydro-dynamic efficiency characteristic of oblique irradiation. PDD is not expected to reach IFE class gains, but is predicted to achieve ignition at energies at around 1 MJ. This is well within the capabilities of the NIF. PDD is also expected to test some of the key direct-drive physics.

IFE High Gain Target designs must be carried out in conjunction with a specific laser, as the target performance depends quite critically on the individual laser characteristics, such as focal profile uniformity, wavelength, and bandwidth. As of this writing, the only IFE driver systems that have had enough development to establish their utility for IFE are the electron beam pumped KrF laser and the frequency tripled DPSSL. Ultraviolet (UV) light is required to maximize coupling to the target and minimize the effects of LPI. KrF laser at 248 nm, whereas the DPSSL, after frequency tripling, provides 351 nm light. The use of an array of fiber lasers has also been proposed, but other than a projected higher beam uniformity, their characteristics are the same as the DPSSL system. DPSSLs have demonstrated an overall efficiency, without the direct drive required frequency conversion, pulse shaping, or beam smoothing, of around 10%. However DPSSL researchers expect that with advances in laser technologies, the efficiency of a fully direct-drive qualified DPSSL would be in the range of 10–15%. KrF lasers are projected to have an efficiency of 7%, based on demonstrated performance of the individual components. So the gain needed for the usual $\eta G > 10$ metric are 67–100 for DPSSL and 140 for KrF. However the previous section showed the importance of high gain, and as shown later in this section, targets driven by KrF lasers are predicted to have significantly higher gain, with more reduced risk from instability than DPSSLs.

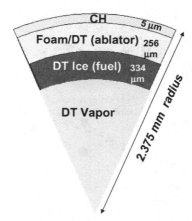

Fig. 6. Compositions and dimensions of a high gain target design. The low density foam is made of carbon and hydrogen and has a mass density of 100 mg/cc.

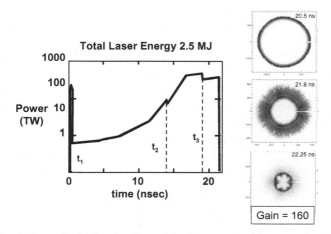

Fig. 7. Simulations of a high gain direct-drive laser target (Courtesy A. J. Schmitt of the US Naval Research Laboratory). The laser pulse shape (intensity vs. time) is shown on the left. 2D simulations of the imploding pellet are shown on the right.

An example of a high gain IFE target is shown in Fig. 6. Figure 7 shows the results of high resolution two-dimensional (2D) simulations of that target. Both are courtesy of A. J Schmitt at The US Naval Research Laboratory (NRL). The simulations include the expected surface finish on the target as well as non-uniformities in the laser focal profile. It is important to account for these as they can provide the seed for the hydrodynamic

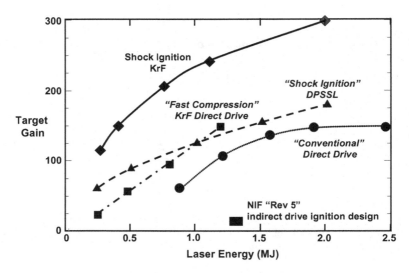

Fig. 8.   Gain curves predicted for various laser direct-drive targets.

instabilities. This design uses a 2.5 MJ KrF laser. The left-hand side of Fig. 7 shows the required laser intensity as a function of time, i.e. the laser pulse shape. The early time high intensity spike is essential to inhibit Richtmyer–Meshkov (instability) growth. The times t_1, t_2, and t_3 indicate times in which the laser spot size is decreased to follow the imploding target. This maximizes laser coupling to the target.

Many institutions have developed codes that have generated the type of simulations shown in Fig. 7. These codes have been benchmarked against experiments on a variety of laser facilities, including the Nike Laser at the US NRL, the Omega and Omega EP Lasers at the University of Rochester, the NIF in Livermore, CA, and the GEKKO XII Laser facility at the Institute of Laser Engineering, Osaka University, Japan.

Figure 8 shows the predicted gain as a function of laser energy for several laser direct-drive target designs: both conventional symmetric drive and shock ignition laser direct-drive targets, with each driven by DPSSL and KrF lasers. These curves are a compilation of one dimensional 1D target simulations from various references, but they have been corroborated with each other. Full 2D simulations have been carried out for several of these designs. For example 2D simulations of shock ignition targets, that incorporate realistic non-uniformities in the target surface or laser, generally show about 65–80% of the gain predicted in 1D. Figure 8 shows that higher performance is predicted for KrF driven targets than for DPSSL driven targets. KrF has a shorter wavelength than DPSSLs ($\lambda = 248$ nm vs. 351 nm), a smoother focal profile, and a broader bandwidth. All of these lead to the predictions of higher gain:

(i) The shorter wavelength of KrF maximizes the coupling efficiency of laser light to target. Hence the drive is more efficient.

(ii) The shorter wavelength also increases the threshold for LPI. Any laser target design (indirect or direct-drive) is subject to LPI. These laser driven instabilities occur in the ablated plasma surrounding the pellet. The pondermotive force (the nonlinear force on a charged particle in an inhomogeneous oscillating electromagnetic field) driving LPI goes roughly as $I\lambda^2$ where I is the intensity and λ is the wavelength. Thus, KrF with its shorter wavelength should be more resistant to LPI than a DPSSL Experiments have born this out. The higher LPI threshold means a KrF target can be driven to higher implosion velocities at lower laser energies, or, with higher pressures and better hydrodynamic stability.

(iii) KrF lasers also produce the smoothest laser beam: a technique called induced spatial incoherence (ISI) produces very high quality focal profiles: less than 0.2% spatial non-uniformities have been demonstrated. That smoothness coupled with the high bandwidth ($\sim$3 THz) minimizes laser induced perturbations, or "imprinting" on the target. A beam smoothing technique called smoothing by spectral dispersion (SSD) has been effectively applied to a DPSSL. While SSD does not produce as smooth a focal profile in all modes as ISI, it should be adequate for most high gain target designs.

In summary, high gain target designs, with sufficient gain for fusion energy, have been generated and the underlying codes have been tested with experiments on existing facilities. KrF lasers are predicted to produce higher gain for direct-drive than a frequency tripled DPSSL. However the best choice for IFE should also be based on factors such as cost and reliability, the achieved overall laser efficiency, and, most importantly, a demonstration of a power plant sized laser beam line.

6 Lasers for Direct-Drive Inertial Fusion

A laser driver for IFE needs to be reasonably efficient, capable of generating multi-kJ of energy in the UV, capable of creating the precise pulse shapes dictated by the target design, and capable of operating with very high reproducibility (1%) for long periods of time (many months) at repetition rates of 5–15 Hz. Presently the e-beam pumped KrF laser and the DPSSL are the primary viable candidates that can meet these requirements for direct-drive, or even indirect-drive laser fusion. KrF is a gas laser that is pumped with high voltage, high current electron beams (500–800 keV, 100–500 kA). The fundamental wavelength is 248 nm. DPSSLs are solid state lasers that are pumped with an array of high efficiency (>60%), high power (>100 kW peak power) diodes. Among the mediums investigated are Yb:S-FAP, the NIF choice of Nd, glass, and cryogenically cooled ceramic Yb:YAG. In all cases the fundamental wavelength is 1051 nm, but DPSSL's can be tripled to 351 nm. KrF lasers for the fusion energy application have been developed primarily with the Electra Laser at the NRL, in Washington DC, whereas the DPSSL for the fusion application has been developed primarily with the Mercury Laser at LLNL Lawrence Livermore National Laboratory, Livermore, CA. Both Electra and Mercury are subscale systems that have developed efficient and durable repetitively pulsed technologies that are scalable to a full size beam line for a power plant. Other KrF laser development for fusion energy has been carried out on single pulse lasers on the Nike Laser at NRL, The Garpun Laser at the Lebedev Institute in Moscow, the Super Ashura Laser at the Electrotechnical Laboratories of MITI in Japan, the Sprite, and Titania Lasers at the Rutherford Appleton Laboratory in the UK, and the Aurora Laser at Los Alamos National Laboratory in New Mexico. Other DPSSL development for fusion has been carried out at the Rutherford Appleton Laboratory, UK, and the HAMA Laser at Osaka University/Hamamatsu in Japan. Smaller KrF Lasers and DPSSLs have been developed for many other industrial and commercial applications.

6.1 *DPSSL laser state-of-the-art*

The DPSSL technology and progress is described in detail in Chapter 4 and will only be briefly summarized here. The Mercury DPSSL has produced greater than 55 Joules of laser light (1051 nm) and has operated continuously for several hours at a repetition rate of 10 Hz. A portion of the output has been converted to 523 nm (frequency doubled) at 160W average power with 73% conversion efficiency using yttrium calcium oxyborate (YCOB). As discussed in the previous section, based on expected advances in the laser materials and needed technologies, a DPSSL based system is projected to have an efficiency of up to 15%. This includes the power needed to cool the

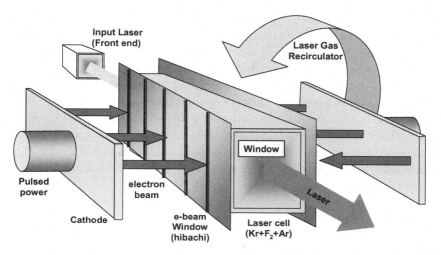

Fig. 9. Components of an electron beam pumped KrF laser system.

crystals. The Mercury Project made many advances including high power diode arrays, forced gas cooling of the laser crystal, high power (200 W) pockels cells, adaptive optics, and high efficiency frequency conversion to the third harmonic. The high power diode lifetime has been demonstrated to be over 140,000,000 shots. A smaller front end for Mercury that uses the same diode/crystal architecture was recently completed and produces 500 mJ (0.5 J). It has run for well over 10 M shots with an RMS stability of 0.78%. An architecture for the DPSSL has been developed that is based on Line Replaceable Units to allow maintenance and replacement during full power operation.

6.2 *KrF laser basics*

KrF is an excited dimer (excimer) gas laser. Discharged pumped KrF lasers have been in routine use for lithography, industrial processing, and medical applications. However large KrF lasers of the size required for fusion energy must be pumped by electron beams. The basic components of an electron beam pumped KrF laser are shown in Fig. 9. The electron beams are emitted from a field emission cathode, driven by a fast pulsed power system. The electron beam propagates through a thin foil into the laser gas. The foil physically separates the electron beam generation (diode) region, which is at vacuum, from the laser cell, which is at atmospheric pressure or above. The structure that supports the foil is called a "hibachi". Two electron beams are injected into the laser cell from opposite sides. The voltage and laser gas pressure are adjusted to give a uniform energy deposition profile across the laser cell. In a repetitively pulsed system a recirculator cools and quiets the laser gas between shots. An external magnetic field prevents the electron beam from self pinching as it propagates through the vacuum diode, and guides it through the laser-gas cell.

A KrF laser is attractive for a power plant because the pulsed power is inherently a robust industrial technology befitting an electric power station, and the gas medium is easy to cool. In fact, in the Sombrero Power Plant study discussed in Section 2, the waste heat from the gas is recovered to boost the overall system efficiency by 2%.

6.3 *KrF laser state of the art*

The Electra KrF laser has run at 2.5–5 Hz and produced between 300 and 700 J in an oscillator mode. Electra has run continuously for 10 hours at 2.5 Hz (90,000 pulses) and 50,000 shots in two runs at 5 Hz. Over 320,000 laser shots were taken in an eight-day period. The continuous run lifetime is now believed to be limited by erosion of the spark-gap-based pulsed power that drives the electron beams. An all solid-state pulsed power system has been built using components that have demonstrated lifetimes in excess of 300,000,000 shots. Figure 10 shows a photo of Electra KrF laser facility.

The system has run continuously at 10 Hz for over 11,500,000 shots (>13 days) and has a pulse to pulse reproducibility of better than 1%. This will become the basis for future KrF pulsed power systems.

The following sections describe the progress in the other KrF components.

6.3.1 *Electron beam emitter (cathode)*

Several types of cathodes have been developed. The best balance between uniformity and durability is an array of carbon fibers that have been pyrolitically bonded into a carbon substrate. This type of cathode has demonstrated over 500,000 shots at 2.5 Hz without any failure.

6.3.2 *Electron beam stability*

Large area, low impedance electron beams are subject to a transit time instability that imparts an axial velocity spread to the electron beam. This lowers the energy transfer efficiency into the laser gas. Modeling with a

Fig. 10. The electra KrF laser facility. The two pulsed power systems are contained in the white tanks flanking the facility. The laser beam output is between the two black magnet coils in the center of the photo.

particle in cell (PIC) code, followed with experiments showed that the instability can be mitigated either by slotting the electron beam cathode and loading the slots with microwave absorbing material, or by making the emitter from a ceramic honeycomb structure.

6.3.3 *Beam transport and deposition*

A method of patterning the electron beam, so it "misses" the ribs of the foil support structure has been demonstrated. This improves the electron beam deposition efficiency into the gas from the range of 35–40% to over 80%. A predictive capability to prescribe the precise topology of the ribs has been developed using a full 3D PIC simulation. This was achieved by Voss Scientific using the large scale plasma (LSP) code. The simulations accurately predict both the cathode counter rotation angle and the energy deposition efficiency. In addition, several 1D codes have been developed to predict the electron beam deposition in the gas. These include transmission and scattering through the foil as well as backscattering in the gas. These have been used to accurately predict the observed energy deposition in the laser gas, as well as in the electron beam window (hibachi foil). The latter is essential for designing foil thermal management systems.

6.3.4 *Electron beam window durability*

As of this writing, the main cause of failures of the electron beam window foil has been identified as post main pulse reversed voltages in the electron beam diode. These reverse voltages can cause the foil to emit electrons in localized areas at very high current densities, which in turn results in localized melting of the foil. Reverse voltages inherent to the pulsed power system design can, to some extent be eliminated by tuning the pulse power. This was shown to extend the foil lifetime. However after long duration runs (~100,000 pulses) the spark gap switches used in the pulsed power system wear and perform erratically, impressing a severe reverse voltage on the diode. This situation will be eliminated by the solid state pulsed power system described above.

6.3.5 *Electron beam window thermal management*

The Georgia Institute of Technology and NRL have developed an efficient technique to effectively cool the pressure foil. A linear array of gas jets is arranged to blow pressurized recycled laser gas directly on the foil. Prototype experiments on the Electra Laser showed the foil temperature during a 5 Hz run was kept at 370°C, which is well below the thermal fatigue limit of 480°C for the stainless steel foil. Just as important, no degradation in the focal profile of a probe laser beam was observed. The power consumption of this system was estimated to be below 2% of the total input power.

6.3.6 *KrF physics code*

NRL has developed a KrF physics code, called Orestes, to design future KrF laser systems. Orestes includes over 22 species, 130 reactions, two excited electronic states of KrF, and 53 vibrational levels. The code accounts for the electron beam input, laser input, plasma thermal and internal energies, the Amplified Spontaneous Emission (ASE), and the laser output. As shown in Fig. 11, Orestes accurately predicts the laser pulse shape for a given electron beam deposition profile. The total laser output in this case was 731 J. The intrinsic efficiency (laser energy out divided by electron beam energy in) during the flat portion of the pulse was projected to be 12%, if high quality windows were used. Orestes also accurately predicts the laser output of several different KrF laser systems over a wide range of conditions. NRL has developed a predictive pulse shaping capability that prescribes how to generate the precise pulse shapes needed for both conventional and shock ignition targets.

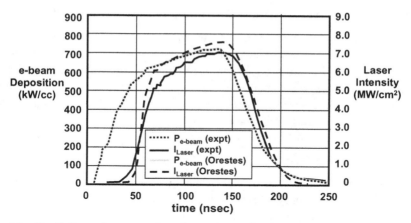

Fig. 11. Predictions of orestes code and experimentally observed laser temporal profiles. No scaling was applied between code and experimental data.

Table 2: Overall Efficiency Expected for a Fusion-Sized KrF Laser System.

Component	Justification/basis	Predicted efficiency (%)
Pulsed power	Demonstrated on all solid state pulsed power	82
Hibachi structure	Demonstrated experiments with no anode foil, patterned e-beam, extrapolated to 800 kV electrons	81
Intrinsic efficiency	KrF literature, demonstrated on Electra	12
Optical train to target	Commercial state of the art optics	95
Ancillaries (pumps, gas circulation, etc.)	Estimated, includes foil and gas thermal management	95
TOTAL		7.1

6.3.7 *Efficiency*

Table 2 shows that the overall efficiency for a fusion reactor sized KrF laser system is expected to be about 7%. These efficiencies are based on experiments and demonstrations to date. Note this is a true "wall plug" efficiency, from electrical input to laser light on target, and includes the power required for thermal management.

7 State of the Art of other Technologies needed for IFE

This section describes the state of development of the other components needed for a Direct-Drive Laser Fusion system. These are the components shown in Fig. 2. In many cases the underlying principles have been demonstrated in sub scale tests. These are not necessarily the optimal or even the only solution, but they have been developed in concert with one another to ensure that there are no interface issues.

7.1 *Final optical train*

The laser light is delivered to the target at chamber center by an optical train. The fusion energy is released by the target as energetic neutrons, ions, and X-rays. The final optic (optical element) will always be exposed to these, but the optical train is designed, through a series of reflectors and labyrinths, to minimize the exposure to the upstream optics. Neutrons cause the most damage, as they displace atoms and cause the optic to swell, thereby compromising the high optical quality needed to precisely illuminate the target. A solution developed at the University of California (UCSD), San Diego, is to make the final optic from a thin, high reflective coating on a silicon carbide or aluminum base. Both are known to be resistant to neutron induced swelling. Orienting the optic at a grazing incidence to the incoming laser beams minimizes the laser fluence on the reflector. The design of the optical train calls for this grazing incidence metal mirror (GIMM) to withstand a laser fluence of 1.5–2 J/cm^2 per pulse. Research has shown a reflector made of a solid solution alloy of aluminum with 5% copper can withstand well above 3.5 J/cm^2 for over 10 million pulses with no damage. Neutrons scattering off the GIMM will be absorbed by the upstream optics. GIMMs are large, so it is desirable that these optics be smaller. Studies at Oak Ridge National Laboratory, in conjunction with Penn State University, showed a tailored high reflectivity, high laser damage threshold dielectric mirror can withstand the expected neutron flux for the 30 year life of the plant. It may be possible to use this as the final optical element.

An alternative to the GIMM plus final optical train is to use two large (1 m diameter) fresnel lenses on either side of a thick shielded neutron "pinhole". Studies by Livermore showed the neutron damage from each pulse will continually anneal if the lens is less than 1 cm thick and kept at 500°C. This approach only works at the DPSSL wavelength of 351 nm, as at 248 nm no annealing was observed.

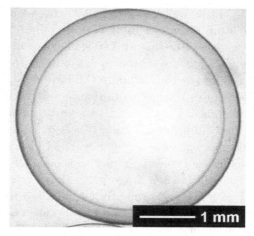

Fig. 12.   Foam shell that meets the requirements of the high gain target shown in Fig. 6.

7.2 *Target fabrication*

The target shown in Fig. 6 is composed of a foam shell with DT wicked into it. General Atomics in California has demonstrated mass production of divinyl benzene foam shells using a triple orifice droplet generator. The shells were produced at 1/min, meet all the dimensions and the density required of the target design, and have a variation of less than 1%. Figure 12 shows a photo of one of those shells.

Other progress in target fabrication includes:

(i) A method to apply an Au–Pd alloy coating on the target. This helps the target physics as it provides a soft X-ray drive during the beginning of the pulse. It also helps target fabrication as it allows fast DT permeation times. Furthermore, the coating provides a reflective layer in the infra-red to both prevent the cryogenic target from warming above melting as it traverses the hot chamber, and allow target tracking by providing a reflective surface for an alignment laser.

(ii) The foam shells need an overcoat of plastic to contain the frozen DT. General Atomics and Schafer Corporation have made leak tight coatings 10 microns thick, but the target specifications call for 2 microns thickness. This will be addressed in future research.

(iii) A cost estimate of $0.16 each for mass production and injection of these targets, based on a chemical engineering analysis of all the process steps. This is well under the $0.25 cost requirement cited by the Sombrero study shown in Table 1.

(iv) Demonstration that ultra-smooth DT ice layers can be grown over a foam underlay, and that smoothness can be maintained at temperatures at least as low as 16 K. A pure DT layer with no foam can not survive temperature swings of more than a few tenths of a degree K. As shown in Section 7.3.3, this ability to start at a lower temperature helps target survival during injection.

(v) An alternative approach, investigated by The University of Rochester, used microfluidics and electrophorisis to sequentially form the foam into shells while still in their liquid phase. Each finished shells would then be filled with DT and promptly injected into the chamber. This "just in time" delivery would forgo the seal coat in favor of relying on precise sublimation of excess DT during injection to produce a target that meets the specifications.

7.3 *Target injection*

7.3.1 *Injector*

A new target can be injected into the chamber only after the blast from the preceding explosion has cleared. At a repetition rate of 5 Hz, this requires a velocity of around 50–400 m/sec, depending on the chamber configuration, dimensions, and chamber recovery time. A direct-drive target is fragile and is at cryogenic temperatures, so an insulating sabot of some kind is required in the injector. One approach is to use a gas gun, in which the sabot is separated and captured before it leaves the muzzle. General Atomics demonstrated this concept with non-cryogenic targets. The system ran at 5 Hz, achieved velocities of 400 m/sec, and demonstrated pointing accuracies of 10 mm. Figure 13 shows the sabot separating at 400 m/sec.

Another approach proposed by AER, Inc is to use an electromagnetic launcher based on the acceleration of a ferromagnetic sabot in a fixed magnetic field gradient. Stability analysis showed this type of launcher would stably accelerate the sabot to the required velocities. The sabot is shaped like a cup and would be rapidly slowed and captured by a reversed magnetic field at the muzzle, thus allowing the target to continue into the chamber.

7.3.2 *Tracking*

General Atomics and UCSD demonstrated a tracking concept that meets the target design specifications for pointing accuracy. The concept is based

STEP 1
Launch

STEP 2
In flight separation

STEP 3
Deflect sabot pieces,

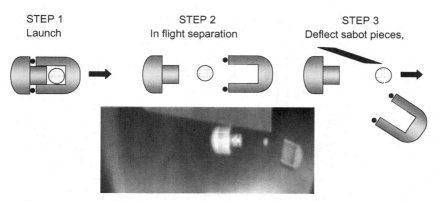

Fig. 13. Concept of separation of the sabot from the target near the muzzle. The sabot is caught within the gas gun and the materials recycled.

on a four stage process: In the first stage, optical sensors precisely determine the target's diameter and launch velocity. In the second stage, which encompasses most of the target's flight, a separate optical sensor continuously monitors lateral deviations of the target's trajectory. In the third stage, which is immediately prior to arrival at the chamber center, the target is illuminated by a short pulse low-intensity "glint" laser. The glint is reflected off the target — recall the IFE target has a reflective Pd–Au coating and is returned backwards through each laser beam line optical train. It is deflected *before* the final steering mirror onto a spatial coincidence detector. At the same time, an alignment laser is directed through the beam line optics and deflected, *after* the final steering mirror, onto the coincidence detector. In the fourth stage, the final steering mirrors direct the alignment laser spot to match the glint laser spot on the detector. Thus the main drive beams can precisely illuminate the target. The glint uniquely defines the position of the target, and can be clearly seen from all angles in one half of the target plane. Thus only two glint lasers are needed. The system was demonstrated using a surrogate target in free fall, and steered a low power laser beam to engage the target with an accuracy of better than 28 microns. At present the high gain target requirement is for an accuracy of 20 microns.

7.3.3 *Survival into the chamber*

The target at cryogenic temperatures must not heat up above the triple point of DT (19.79 K) as it is injected into a chamber. As discussed below,

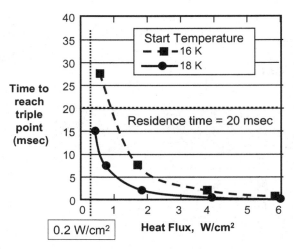

Fig. 14. Calculation of the time to reach the triple point as a function of the heat load and the initial target temperature.

all the selected chamber designs have a near vacuum background. Thus the only heat load on the target is from the wall at about 1,000 K, or about 0.2 W/cm^2. At first glance, getting the target to survive this journey into the chamber looks quite daunting, but simulations, experiments, and advances in target fabrication make this feasible. Figure 14 shows a calculation by UCSD of the time it takes the target to reach the triple point as a function of the heat load and the initial target temperature. The target surface is expected to have an infrared emissivity of 0.95, in accordance with bench tests on the Au–Pd coating.

The residence time of the target in the chamber is 20 m/sec, as shown by the horizontal dotted line. The anticipated heat load on the target is shown by the vertical dotted line. From Fig. 14, if the target initial temperature is 16 K, it will not exceed the triple point during injection. The calculations in Fig. 14 were in agreement with experiments performed at Los Alamos National Laboratory, in which a sudden heat flux (60% of the prototypical value) was applied to a DT ice layer. The initial target temperature was 19 K and had no foam. The layer did not degrade for over 20 m/sec. Thus it is expected that an IFE target, in which the DT is wicked into a foam shell, and thus capable of starting at temperatures as low as 16 K, should survive injection into the chamber.

7.4 *The reaction chamber*

The reaction chamber is one of the most challenging aspects in developing any practical fusion power plant. It is no different for direct-drive laser fusion. The six functions of any IFE chamber are to:

(i) Ensure the target survives injection and/or placement (i.e. does not deform or melt).

(ii) Allow the driver energy to be coupled to the target with the required precision.

(iii) Repeatedly withstand the pulsed emissions of X-rays, ions, and neutrons from the fusion reaction.

(iv) Recover conditions to allow the next "shot".

(v) Provide a means to convert heat to electrical power.

(vi) Provide a means to breed and recover tritium.

The chamber and target design are dependent on each other, as each target design produces a distinct "threat spectrum," (energy partitioning among the reaction products) that must be withstood by the chamber's first wall. In an IFE target, about 75% of the fusion energy is incident on the wall as neutrons. The rest of the energy is in X-rays and ions. The relative energy between X-rays and ions, as well as the spectrum of the ions, varies significantly between indirect and direct-drive targets. Note all IFE targets produce the usual energy partitioning between neutrons, ions and X-rays characteristic of a fusion reaction. However some of this energy is absorbed by the surrounding dense cold shell. This reduces the energy of the primary products and creates an energetic spectrum of secondary ions and X-rays. In an indirect-drive target the vaporized hohraum mass absorbs almost all the ion energy and converts it to X-rays.

Table 3 shows the energy partitioning, 100 nsec after burn, of all the constituents of a high gain conventional direct-drive target. About 1.3% of the energy is released in X-rays, 24% in ions, and the remainder in neutrons. Not shown in the table is that these ions have a significant spread in velocity. This ratio is representative of all direct-drive target designs, including shock ignition. How these emissions deposit energy into the first wall is important for the chamber design: the X-rays arrive first, as they propagate at the speed of light. Their typical energy is a few keV, so they deposit that energy within the first micron of the wall's surface and on a time scale of a few nsec. The ions are slower, and arrive several microseconds (μ sec) after the X-rays. Because of their energy spread, they deposit their energy over

Table 3: Energy Accounting from a High Gain
Target (Laser 2.46 MJ, Gain 150).

Species	Total (J)	Fraction of total energy (%)
X-rays	4.937×10^6	1.34
Gammas	1.680×10^4	0.0045
Neutrons	2.743×10^8	75
Protons	1.763×10^7	
Deuterons	2.172×10^7	
Tritons	2.650×10^7	
^{3}He	7.777×10^4	
^{4}He	3.028×10^7	
^{12}C	6.880×10^6	
^{13}C	8.367×10^4	
Pd	1.844×10^5	
Au	3.607×10^5	
Total ions		23.65
TOTALS	3.671×10^8	100

$\sim 2\,\mu$sec. More importantly, they deposit that energy within the first $5\,\mu$m of the surface. The neutrons deposit their energy volumetrically.

As an example of how the threat spectrum heats the first wall, consider a 200 MJ (output) fusion target, an evacuated chamber with a radius of 8 m, and as discussed below, a tungsten-clad steel wall facing the target. The initial temperature of the tungsten is 800°C. The X-rays carry enough energy, and have a short enough deposition depth, to heat the wall by 800°C. So the surface heats to 1600°C. Tungsten conducts heat sufficiently to cool the surface back to 800°C before the ions arrive several microseconds later. The ions deposit their energy over 2–3 msec, and even though some heat is conducted away during the pulse, the surface is re-heated by 1600–2400°C. Nevertheless, this is comfortably below the tungsten melting point of 3400°C.

For indirect drive, the energy partitioning is roughly inverted from Table 3 to about 20% X-rays, and 5% in ions. Consider the effect this has on the same chamber as above. The X-rays heat the tungsten surface by about $(20/1.3) \times 800 = 12,300$°C. This is enough turn the tungsten surface into a plasma. This analysis is a bit simplistic, and conditions can be adjusted to lower the temperature somewhat. But this is not enough. There is just no mortal material that can survive that onslaught of X-rays. Thus any indirect-drive target requires some continuously replenishable buffer or replaceable wall to protect the underlying solid structure. This can be a thick liquid (metal or molten salt), as chosen for the heavy ion or Z-pinch approaches, or a buffer gas, as chosen for the LIFE laser indirect-drive concept. Direct drive, however, offers broader choices.

Table 4 shows the chamber first wall protection schemes that have been considered for direct drive. While all are potentially workable, having a vacuum in the chamber is currently the most attractive option because simulations show there are no issues with either target injection/survival or high fidelity laser propagation to the target. In addition, it should be possible to test the key physics unknowns without building a full scale fusion system. The two most developed of the vacuum chamber concepts are to use a Solid Wall or Magnetic Intervention.

7.4.1 *Solid wall chamber*

For solid wall chamber concepts, the first wall was chosen to be a thin ($\sim$1 mm) tungsten layer bonded to a low activation ferritic steel substrate. This segregates the tungsten layer, which serves as an armor against the target emissions, from the steel, which carries out the structural functions. The thickness of the tungsten was chosen to be thick enough to smooth the cyclic thermal stresses at the tungsten/steel interface, yet thin enough to ensure adequate heat removal before the next pulse. A number of experimental facilities have been employed to simulate the thermal cycle and particle deposition: The RHEPP facility at Sandia Laboratory provided a source of repetitively pulsed high energy ions, a high intensity infrared lamp at Oak Ridge National Laboratory mimics the cyclic heat loading at the interface, and the Dragonfire laser facility at UCSD was used to precisely study the evolution of long term cyclic heating of the armor. The experiments were supported with the Unified Materials Response Code under development at UCLA and the University of Wisconsin. The experiments and modeling suggested that below 2,400°C there is little evidence of long term mass loss and no long term damage at the tungsten/steel interface. The 2400°C limit can be met with a 350 MJ direct-drive target in an evacuated chamber with a radius of 11m.

The main issue that is still unresolved is helium retention: high energy (several MeV) helium ions produced by the target are driven 2–5 microns deep into the surface. Because the helium migration distance is only 50–100 nm, the helium coalesces into bubbles and at grain boundaries. The resulting build up in helium pressure exfoliates the surface. Experiments at Oak Ridge National Laboratory and the University of North Carolina

Table 4: Chamber First Wall Protection Schemes Considered for Direct Drive.

Concept (wall/chamber)	Advantages	Challenges
Solid wall-vacuum	• No chamber recovery issues • No laser propagation issues • No target injection issues	• Materials issues
Magnetic intervention-vacuum	• Smallest chamber • Minimal thermal load on first wall • Eliminates He retention issues • No laser propagation issues • No target injection issues	• Ion dumps
Replaceable solid wall-vacuum	• No laser propagation issues • No target injection issues	• Operational complexity • Potential waste issues
Solid wall-gas in chamber	• Smaller chamber	• Laser propagation • Target survival and placement • Chamber recovery (hot gas and residual plasma) • Gas inventory/gas handling
Thick liquid walls	• No materials issues • No neutron damage	• Chamber recovery • Laser propagation in droplets/mist • Target survival through droplets/mist • Difficult to develop or modify

showed that the problem is not as severe in the solid wall chamber, because the natural IFE cycle of pulsed implantation followed by a high temperature pulse can enhance the helium diffusion. Another solution that was conceived by UCSD and tested in a small scale electrostatic trap at the University of Wisconsin was to make the first wall of a fibrous tungsten material whose characteristic length is smaller than the migration distance. Thus, no matter where the helium is stopped, it is always less than a migration length from a free surface. Results from the first experiments showed the expected mass loss from an IFE sized chamber would be only a 1-μm layer of solid material over the equivalent of 450 days of operation.

7.4.2 *Magnetic intervention*

From the above discussion, the ions are the most problematic from the point of view of designing a robust first wall. However the ions are charged, and hence can be deflected by magnetic fields. If a cusp magnetic field is applied to a chamber, the ions will be deflected through poloidal holes and an external belt. The ion energy could then be absorbed in external dumps. This isolates the chamber functions of energy absorption from those of target injection and laser propagation. The concept was first demonstrated in the 1970's by an experiment at NRL. More recently that experiment was accurately modeled by Voss Scientific in New Mexico. The ion energy can be dumped in a flowing metal "waterfall" or metal vapor. Gallium is a good choice because of its low melting temperature and extremely low vapor pressure (10^{-6} Torr at 720°C). Magnetic Intervention allows for a smaller chamber, which not only lowers cost, but alleviates the velocities required for target injection. There is also more flexibility in the first wall, because the only heat source is the X-rays. By going to a lower Z material, the X-ray penetration depth is increased, the peak wall temperature is substantially decreased, and the cyclic thermal fatigue is reduced. For example, with a silicon carbide first wall, the increase is expected to be about 140°C. One concept for a Magnetic Intervention Chamber is shown in Fig. 15.

7.5 *Chamber breeding, tritium handling and power management*

Outside the first wall of the chamber, an IFE system can adapt most of the concepts developed for a magnetic fusion power plant. Several blanket designs have been produced for both magnetic intervention and conventional chambers. These designs use either a molten salt composed of fluorine, lithium, and beryllium (FLIBE) or liquid lead-lithium (PbLi) as the breeder material. Three power conversion concepts have been evaluated, all based on a three compressor/two intercooler Brayton Cycle. One solid wall concept uses amore conventional reduced activation F82H steel wall

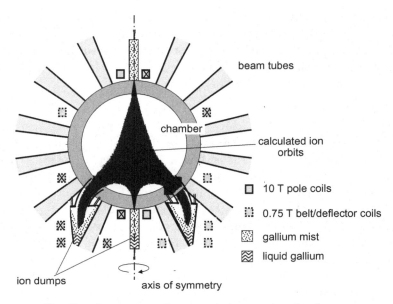

Fig. 15.   One concept of a magnetic intervention chamber.

and tungsten armor. The initial chamber temperature is 550°C, and the electrical conversion efficiency is estimated to be 38%. A second solid wall concept uses a more advanced Ferritic Steel wall and tungsten armor. The chamber operates at 700°C, and the electrical conversion efficiency can be as high as 49%. For a magnetic intervention chamber concept, the first wall is a SiC/SiC composite, which runs at 1,000°C and the conversion efficiency is predicted to be 50%.

8 A Plan to Develop Laser Direct-Drive Fusion

The NRL and its collaborators have proposed a plan to develop Direct-Drive Laser Fusion energy based on the KrF laser. There may be similar plans developed by others, but this is the only one that has been published in the archival literature. While this was developed for a KrF laser, it could be adapted for the DPSSL approach with a higher laser energy. The centerpiece of the program is development of *The Fusion Test Facility* (*FTF*), which would demonstrate that all the components work together with the required precision, durability and efficiency. This includes tritium breeding. The FTF would be large enough (150 MW fusion power) to fully test materials and components for the power plants. Most importantly, it would allow a high confidence capability to both predict the system performance, and to ascertain the viability of this concept.

8.1 *The three phases of the development plan*

8.1.1 *Phase I: Develop full scale components*

(i) Develop and test a full scale laser beam line (16–28 kJ, 5 Hz,) that meets the fusion requirements for efficiency, durability pulse shape, and beam smoothness.

(ii) Demonstrate the full scale beamline which engages injected targets with the required precision and repetition rate.

(iii) Refine KrF target physics at high energies using the full scale KrF beamline. This would include direct-drive target implosion physics on the NIF.

(iv) Develop the needed technologies for the chamber, optics, low cost target fabrication, and materials to the level needed for the FTF.

8.1.2 *Phase II: The fusion test facility*

(i) Demonstrate that all the components work together with the required precision, durability, and efficiency.

(ii) Produce around 150 MW of fusion power.

(iii) Serve as a platform to develop and test first wall and optics materials, as well as concepts for the chamber, tritium breeding, and thermal management.

(iv) Fully test the effects of fusion neutrons on materials and subscale components for fusion power plants.

(v) Provide an economical, technical, and operational basis for power plants.

8.1.3 *Phase III: Build a pilot power plant*

(i) Provide electricity to the grid.

8.2 *Description of the FTF*

A conceptual design for the FTF is shown in Fig. 16. The FTF concept shown in Fig. 16 would have 20 28 kJ KrF amplifiers, each producing 96, 2.5 nsec long, angularly multiplexed beams. In a more recent concept, the FTF would have 35 amplifiers, each producing 16 kJ. In either case, the

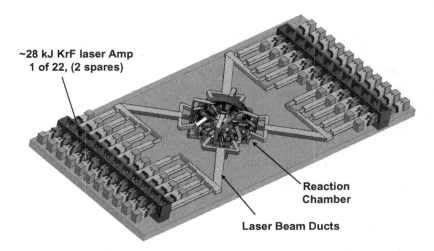

Fig. 16.   One conceptual design of the FTF.

amplifier size would be sufficient for a full scale reactor beam line. The beams from these amplifiers are grouped into 40 clusters that uniformly illuminate the target.

8.2.1 *FTF target designs*

The FTF target designs would be based on either the KrF Fast Compression or the KrF Shock Ignition gain curves shown in Fig. 8. The predicted 1D gains with a 500 kJ KrF laser are 75 and 160, respectively. Allowing for a degradation of 80% for 2D effects, as seen in simulations, at 5 Hz these correspond to a thermal output power of 150 MW and 320 MW.

8.2.2 *FTF laser*

Each amplifier/beam line in the FTF laser would be a modest extension in size of the present Nike amplifier, but would incorporate the advances in efficiency and durability made on the Electra facility. The main Nike amplifier has a 60 cm × 60 cm square optical aperture, a 100 cm long cathode, and is pumped by twin opposing 600 kV electron beams. The output of Nike is 5 kJ, but the Orestes Code simulations described in Section 4 suggest it can produce as much as 8 kJ by incorporating the efficient e-beam technologies developed on Electra. For the FTF amplifier, the aperture width would be increased to 100 cm, the cathode length to 250 cm, and the voltage increased from 600 to 800 kV. The higher voltage would both accommodate the wider gas cell and reduce losses due to electron beam deposition in the pressure foil. Orestes simulations predict a little more than 17 kJ output.

8.2.3 *The other components of the FTF and tritium breeding*

The other components of the FTF will be based on the technologies discussed in Section 7. Tritium breeding will be required to ensure a sufficient supply of fuel. For a 150 MW facility running at 60% availability, the tritium requirement would be about 5 kg per year, assuming the unburnt tritium is recovered. This is a significant fraction of the estimated tritium inventory currently available. Thus the FTF must breed in order to have sufficient tritium available. The two blanket concepts described in Section 7 would be appropriate for this. For example, the self-cooled PbLi blanket concept would give a tritium breeding ratio (TBR) = 1.1, accounting for losses in the beam ports. If an additional 9% of the solid angle were blocked for materials tests, as described below, then the TBR would fall to 1.0. The FTF could also evaluate several blanket concepts by adopting the "Test Blanket Module" concept pursued by ITER. The FTF could be configured as an electricity generator, not necessarily to be connected to the grid, but to demonstrate all the power conversion technologies and to provide enough electricity to run itself.

8.2.4 *The FTF as a material and component development platform*

The FTF would be a source for high flux, high energy neutrons to test and develop candidate reactor materials and components. As no such capability exists, this would be invaluable to both the MFE and IFE communities. There is ample room to place materials and even subscale components inside the FTF chamber because the neutrons come from a point source. The FTF chamber wall can be made comfortably large (5 m in radius) to ensure that the chamber itself is not the material being tested. For example, an object placed 1 m from the target can be exposed to up to 50 dpa/year, (dpa = displacements per atom) yet the chamber wall would be exposed to only 2 dpa per year. 50 dpa is about five times that expected for the components in a first generation fusion reactor. These numbers assume 150 MW of fusion power and 60% availability. Single objects with volumes as large as 400 L can be tested in the FTF chamber. Once material and component issues are resolved, one could consider redirecting the laser beams into a smaller diameter chamber where the neutron fluxes at the wall match the levels desired for a power plant.

8.3 *Timescale to deliver direct-drive IFE*

With suitable resources, Phase I should take about five years, and the FTF would be ready for testing in another five. Allowing for five years for FTF operations, it is not unreasonable to expect construction of a pilot power plant to deliver electricity to the grid to be started in the early 2030s. The cost of the FTF has not been established. But it is a full nuclear facility, so it is expected the cost will be comparable to that of a small fission power plant.

9 Other Approaches to IFE

This section describes briefly the three other main approaches to developing fusion energy through inertial confinement. Further information can be found in the reading list at the end of this chapter.

9.1 *Heavy ion fusion*

Heavy ion targets are similar to laser targets, but are driven by pulses of high energy (GeV range) heavy ions. Heavy Ion Fusion research has been carried out by the University of California Berkeley, Princeton Plasma Physics Laboratory in New Jersey, GSI in Darmstadt, Germany, and the Institute for Theoretical and Experimental Physics in Moscow.

The principal reason for evaluating Heavy Ion Fusion is that high-energy particle accelerators producing megajoule-scale beam energies have separately exhibited the pulse rates, average power levels, and durability required for IFE. The expected relatively high efficiency of a heavy ion driver permits the use of indirect-drive and still exceed the $\eta G > 10$ parameter. As discussed in Section 7, heavy ions must use liquid walls to protect the chamber, but that is readily adaptable to this approach.

A wide range of reactor concepts have been explored, including direct and indirect-drive targets, various ion species, and multiple accelerator options. For the accelerator, the two most promising ones are induction accelerators and radio-frequency (RF) accelerators. Many of the reactor studies in Table 1 were also carried out in a companion effort for heavy ion drivers. One conceptual design for a heavy ion power plant uses an induction accelerator, ballistic neutralized focusing, a thick liquid protected wall, and an indirectly driven target. The drive was supplied by singly ionized bismuth ions accelerated to 4 GeV, with a total energy of 7 MJ. The total accelerator length is 3 km. Target gains of greater than 60 were predicted.

More recent calculations indicate it would be possible to achieve gains on the order of 90 to 130 at beam energies from 1.8 to 3.3 MJ, respectively.

9.2 *Pulsed power driven fusion*

Pulsed-power-driven fusion utilizes a fast (100 nsec), high current pulse (50–80 MA), to create a plasma at fusion conditions. Two main concepts have been explored. In an earlier approach, the current ionizes a cylindrical array of wires, which collapses rapidly on axis due to the very high self magnetic fields. The sudden transfer of energy from radial motion to heat produces copious X-rays. Soft X-rays in the range of 0.1–10 keV have been produced with a peak power up to 300 terawatt (TW). These X-rays could then be directed to compress a DT pellet. In the present approach, called MagLIF, the current is made to pass through a cylindrical liner which collapses and compresses a magnetized, and pre-ionized plasma. The plasma is created and preheated with an intense laser. Magnetization helps confine the plasma during the implosion. The liner collapses rapidly to compress and heat the internal plasma to fusion conditions. The advantages of all pulsed power approaches are the relatively low-cost and high-efficiency of the driver technology. If the physics scales to higher currents, scaling the pulsed power to fusion level currents appears to be achievable. At the present these pulsed power approaches are being pursued only as part of the inertial confinement fusion (ICF) program, which only requires single pulse events. Transforming these concepts into a viable fusion energy source will require considerable development to handle the enormous pressures imposed on the surrounding structure. For example, as the energy released in a single pulse can be as high as a GJ, most of the current feeds and surrounding structure would vaporize on each pulse. The repetition rate is rather low, about 0.1 Hz. This offers the opportunity of replacing or recycling the transmission lines (current feeds) between pulses. Most research on pulsed power driven ICF is being carried out by Sandia National Laboratories in New Mexico, but there are also activities in Russia.

9.3 *Fast ignition*

Fast ignition is similar in concept to Shock Ignition in that the objective is to separate the compression step from the ignition step in a conventional ICF implosion. The spherical target is compressed at a relatively slow implosion velocity (200–250 km/sec) to a relatively modest compression ratio. At peak compression, a high energy (50–150 J) petawatt (10^{15} W) laser beam illuminates the target at one point. The laser produces a beam of focused ions or electrons (with currents exceeding 100 MA), towards the center, where it ignites the hot spot. The ensuing burn wave then burns the fuel in the surrounding shell. Fast ignition applies to any IFE driver technology: lasers or heavy ions, direct-drive or indirect-drive, and some pulsed power based approaches. The main advantage of fast ignition is that the fuel does not have to be uniformly compressed. All that matters is that there be a dense enough region for ignition to take place. Until Shock Ignition was seriously studied, Fast Ignition also had the promise of the highest gains of any IFE concept. However Shock Ignition is predicted to have comparable gains without the complication of a second, and technically challenging, petawatt-class laser. Currently Fast Ignition is being vigorously evaluated with the Laser Inertial Fusion Test (LIFT) at The Institute for Laser Engineering (ILE) at The University of Osaka, the HiPER Program of the European Union, and numerous universities.

10 The Path Forward

As shown in this chapter, many advances have been made in development of the science and technologies needed for a practical fusion energy source based on direct-drive and inertial confinement. These advances have been made not just on paper (or, more fittingly, the computer monitor) but in scalable demonstrations in the laboratory. First and foremost, state of the art computer simulations, whose underlying physics has been backed with real experiments, predict more than enough performance for a power plant. "More than enough performance", means there should be ample head room so when a full size system gets built, and Mother Nature has her say, the system will still perform well enough to be an attractive power plant. Two totally different laser technologies have been developed. They both show the potential for meeting the requirements for the fusion energy requirements for repetition rate, durability and efficiency. And there are credible approaches for the other key components such as target fabrication and injection, the final optics, and chamber wall. In many cases the principles have been tested on the bench. Sufficient progress has been made to warrant embarking on the three phase program discussed in Section 8.1 of this chapter. Unfortunately, the current environment, in which the NIF (based on the riskier and more complex lower performance indirect-drive approach) has so far failed to achieve any practical definition of ignition, and in which the ITER effort is consuming most of the resources available for fusion, makes it difficult to get support or such a program. Nevertheless, the scientific and technological achievements for direct-drive inertial fusion are remarkable, and they certainly warrant revisiting this approach to a clean, plentiful energy source for humankind.

11 Glossary

Berkeley	University of California, Berkeley
Blanket	Lithium bearing fluid that breeds tritium and carries a heat to energy conversion device
CBET	Cross beam energy transfer
Conventional Direct Drive	Array of lasers symmetrically compresses fuel and ignites hot spot
Deuterium (D)	isotope of hydrogen with one proton and one neutron
Direct Drive	Laser illuminates the pellet directly
DPSSL	diode pumped solid state laser
DT	Deuterium-Tritium
Electra	KrF laser system at NRL
Fast Ignition	a petawatt laser to ignite hot spot in pre compressed fuel
FTF	Fusion Test Facility concept proposed by NRL and its collaborators
GEKKO XII	Direct drive glass laser facility at ILE, University of Osaka
General Atomics (GA)	Private Company in San Diego, CA
HAPL	High Average Power Laser Program
Heavy Ion Fusion	beams of heavy ions compress fusion pellet. Can be direct or indirect-drive.
HiPER	European IFE concept based on fas ignition or direct-drive
ICF	Inertial Confinement Fusion (The physics behind IFE)
IFE	Inertial Fusion Energy
ILE	Institute for Laser Engineering, Osaka University (Japan)
Indirect Drive	laser heats the inside of the hohlraum hot enough to produce X-rays that drive the pellet

keV	kilo electron volt (one eV = 11,600°K)
KrF	Krypton fluoride laser, electron beam pumped gas laser
LIFE	LLNL proposed concept for indirect-drive laser fusion
LIFT	Laser inertial fusion test, ILE fast ignition inertial fusion power plant design
LLE	Laboratory for Laser Energetics, Rochester New York
LLNL	Lawrence Livermore National Laboratory, Livermore, CA
LPI	Laser plasma interaction
MagLIF	Pulsed power fusion concept: A high current is driven through a liner, inducing it to implode on a pre-ionized, pre-magnetized fuel
Mercury	DPSSL at LLNL
MFE	magnetic fusion energy microsecond (μ sec) = 1 millionth of a second millisecond (msec) = 1 thousandth of a second
MJ	megajoule
MW	megawatt nanosecond (nsec) = 1 billionth of a second
NIF	National Ignition Facility-1.8 MJ laser system at LLNL
NIKE	KrF laser system at NRL
NRL	Naval Research Laboratory (USA)
OMEGA	Laser system at LLE
PDD	polar direct drive
Pellet	Pea sized sphere that contains an outer ablator, and an inner layer of frozen DT fuel. Also referred to as the "target."
Pulsed Power Fusion	See Z-Pinch Fusion or MagLIF
RHEPP	Repetivie high energy pulsed power ion beam facility at Sandia National Laboratory.
Rutherford Appleton Laboratory	Major laboratory in the UK
Schafer Corporation	Private Company, Livermore, CA
Shock Ignition	A strong shock ignites hot spot in pre compressed fuel

SNL	Sandia National Laboratory, Albuquerque, NM
Tritium (T)	Isotope of hydrogen with one proton and two neutrons
UCSD	University of California, San Diego, CA
Z-Pinch Fusion	High currents are driven through an array of wires to produce X-rays that implode a fusion pellet

Acknowledgment

This work has been sponsored by the National Nuclear Security Administration of the US Department of Energy.

Suggested Further Reading

History and power plant studies

M. J. Lubin and A. P. Fraas (1971). Fusion by laser. *Scientific American*, **224**, 21–33.

L. M. Waganer (1994). Innovation leads the way to attractive inertial fusion energy reactors–Prometheus-L and Prometheus-H. *Fusion Engineering and Design*, **25**, 125–143, DOI:10.1016/0920-3796(94)90059-0.

W. R. Meier and C. W. von Rosenberg, Jr (1992). Economic modeling and parametric studies for SOMBRERO — A laser-driven IFE power plant. *Fus. Technol.* **21**, 1552–1556, and references therein.

Direct Drive Target Physics and High Gain Target Designs

A. Basic Book on Inertial confinement

S. Atzeni and J. Meyer-Ter Vehn (2004). *The Physics of Inertial Fusion*. Oxford Press.

B. High gain target designs

S. E Bodner, D. G. Colombant *et al.* (1998). New high gain target design for a laser fusion power plant. *Phys. Plasmas*, **5**, 1901.

A. J. Schmitt, D. G. Colombant *et al.* (2004). Large-scale high-resolution simulations of high gain direct-drive inertial confinement fusion targets. *Phys. Plasmas*, **11**, 2716.

P. W. McKenty, V. N. Goncharov *et al.* (2001). Analysis of a direct-drive ignition capsule designed for the national ignition facility. *Phys. Plasmas*, **8**, 2315.

C. Shock Ignition

R. Betti, C. D. Zhou *et al.* (2007). Shock ignition of thermonuclear fuel with high areal density. *Phys. Rev. Lett.*, **98**, 155001.

L. J. Perkins, R. Betti (2009). Shock ignition: A new approach to high gain inertial confinement fusion on the national ignition facility. *Phys. Rev. Lett*, **103**, 045004.

A. Atzeni, X. Ribeyre *et al.* (2014). Shock ignition of thermonuclear fuel: Principles and modeling. *Nucl. Fusion*, **54**, 054008.

W. Theobald, R. Betti *et al.* (2008). Initial experiments on the shock-ignition inertial confinement fusion concept. *Phys Plasma*, **15**, 056306.

Lasers-DPSSL and Fiber Lasers

A. Bayramian, S. Aceves *et al.* (2011). Compact, efficient laser systems required for laser inertial fusion energy. *Fus. Sci. & Tech.*, **60**, 28–48.

A. J. Bayramian, P. Armstrong *et al.* (2007). The mercury project: A high average power, gas-cooled laser for inertial fusion energy development. *Fus. Sci. & Tech.*, **52**, 383–387.

P. D. Mason, K. Ertel *et al.* (2011). Optimised design for a 1 kJ diode pumped solid state laser system. *Proc. of SPIE*, **8080** 80801X-1.

D. Eimerl, E. M. Campbell *et al.* (2014). Star driver: A flexible laser driver for inertial confinement fusion and high energy density physics. *J Fusion Energ* DOI 10.1007/s10894-014-9697-2.

Lasers KrF

P. M. Burns, M. Myers *et al.* (2009). Electra: An electron beam pumped KrF reprate laser system for inertial fusion energy. *Fus. Sci. Technol.*, **56**, 346–351.

J. D. Sethian, M. C. Myers *et al.* (2004). Electron beam pumped krypton fluoride lasers for fusion energy. *Proceedings of the IEEE*, **92**, 1043–1056.

Other key technologies needed for laser IFE

J. D. Sethian, M. Friedman *et al.* (2003). Fusion energy with lasers, direct-drive targets, and dry wall chambers. *Nucl. Fusion*, **43**, 1693.

J. D. Sethian, D. G. Colombant *et al.* (2010). The science and technologies for fusion energy with lasers and direct-drive targets. *IEEE Trans on Plasma Sci*, **38**, 690–703, and references therein.

J. F. Latkowski *et al.* (2003). Fused silica final optics for inertial fusion energy: Radiation studies and system-level analysis. *Fus. Sci. Tech.*, **43**, 540–558.

A. K. Tucker-Schwartz, Z. Bei *et al.* (2010). Polymerization of electric field-centered double emulsion droplets to create polyacrylate shells. *Langmuir*, **26**, 18606–18611, and references therein.

Development Path

S. P. Obenschain, J. D. Sethian, A. J. Schmitt (2009). A laser based fusion test facility. *Fus. Sci. Tech.*, **56**, 594–603.

A. J, Schmitt, J. W. Bates *et al.* (2009). Direct drive fusion energy shock ignition designs for sub-MJ lasers. *Fus. Sci & Tech.*, **56**, 377–383.

S. P. Obenschain, D. G. Colombant *et al.* (2006). Pathway to a lower cost high repetition rate ignition facility. *Physics of Plasmas*, **13**, 056320.

A. Ying, M. Abdou *et al.* (2006). An overview of US ITER test blanket module program. *Fusion Engineering and Design*, **81** 433–441, and references therein.

Other approaches to inertial fusion

A. Heavy ions

S. S. Yu, W. R. Meier *et al.* (2003). An updated point design for heavy ion fusion. *Fus. Sci. Tech.*, **44**, 266–273.

D. Callahan-Miller and M. Tabak (2000). Progress in target physics and design for heavy ion fusion. *Physics of Plasmas*, **7**, 2083–2091.

B. Pulsed power

T. W. L Sanford, G. O. Allshouse *et al.* (1996). Improved symmetry greatly increases X-ray power from wire-array Z-pinches. *Phys. Rev. Lett.*, **77**, 5063–5066.

S. A Slutz, M. C. Herrmann *et al.* (2010). Pulsed-power-driven cylindrical liner implosions of laser preheated fuel magnetized with an axial field. *Phys of Plasmas*, **17**, 056303.

S. A. Slutz, C. L. Olson, and P. Peterson (2003). Low mass recyclable transmission lines for Z-pinch driven inertial fusion. *Phys of Plasmas*, **10**, 429–437.

C. Fast ignition

M. Tabak, J. Hammer *et al.* (1994). Ignition and high gain with ultrapowerful lasers. *Physics of Plasmas*, **1**, 1626–1634.

G. A. Mourou, C. P. J. Barty and M. D. Perry (1998). Ultrahigh-intensity lasers: physics of the extreme on the tabletop. *Physics Today*, **51**, No. 1, January, 22–28.

Chapter 6

Magnetic Fusion Energy

M. C. Zarnstorff* and R. J. Goldston[†]
Princeton Plasma Physics Laboratory
PO Box 451, Princeton, NJ 08540, USA
* zarnstor@pppl.gov
[†] *rgoldston@pppl.gov*

Fusion of light nuclei confined by magnetic fields is being developed as an effectively inexhaustible energy source without the production of greenhouse gases or long-lived radioactive waste and without the risk of runaway accidents. This chapter presents the characteristics of magnetically confined fusion and methods to harness it as an energy source, including an overview of fuel resources, waste, safety, and proliferation risks. Section 2 discusses the physics and technology of magnetic fusion systems. Section 3 describes current major experiments, fusion energy achievements, and the role of the ITER burning-plasma experiment. Section 4 summarizes the international activities to develop fusion as an attractive energy production system.

1 Overview

The fusion of light nuclei powers the stars and our sun is the primary origin of almost all terrestrial energy. Controlled use of fusion as a direct human energy source has been pursued for more than 60 years because it offers an effectively inexhaustible energy source that would be available to all nations without production of greenhouse gases or long-lived radioactive waste, without the possibility of catastrophic runaway accidents, and with low weapons proliferation risks. In addition, fusion offers a steady source of base load power that should be able to be located anywhere, without need for large amounts of land or energy storage. The challenge of fusion is to confine and control extremely hot, ionized gases called plasma, and this has driven an increased understanding of plasma physics, the study of ionized gasses that make up almost all the visible universe. The increased understanding and development of fusion-related technology is resulting in rapid progress toward providing abundant fusion energy through worldwide collaboration. The next major step will be taken by ITER, an international experiment under construction near Marseille, France, to study burning-plasma conditions at nearly reactor scale. Beyond ITER, the challenges will be to make fusion energy production practical, reliable, and cost-effective.

The easiest fusion reaction for use in energy production is between deuterium and tritium (heavy isotopes of hydrogen) creating helium and neutrons. Tritium is radioactive, decaying with a half-life of 12.3 years and thus must be manufactured. The helium ion produced from the fusion of deuterium and tritium carries 20% of the released energy and heats the plasma to sustain the temperature $T \sim 160\,\mathrm{M^\circ C}$ needed for the peak reaction rate at a given pressure (see Fig. 1). A blanket-structure containing lithium surrounds the plasma absorbing the neutrons, capturing the neutron's energy thermally and producing tritium fuel. A fusion power plant will convert thermal power to electrical power using turbine generators, similar to other

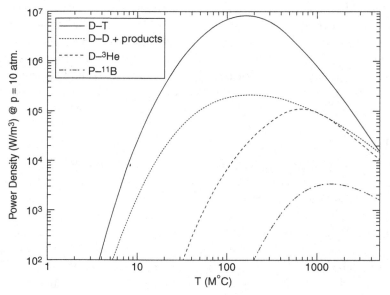

Fig. 1.   Fusion power density vs. fuel temperature for different fusion reactions. Fuel pressure is fixed at 10 atm.

power generation systems, with a conversion efficiency expected to be in the range from 1/3–1/2 depending on blanket temperature. The production of 1 GW-year $(3.2 \times 10^{16}$ J) of fusion electrical energy with a thermal conversion efficiency of 1/3 requires 88 kg of deuterium and 263 kg of lithium-6 (^{6}Li), producing 350 kg of helium. Higher thermal conversion efficiencies, due to blankets being operated at a higher temperature, reduce the required fuel quantities.

At the very high temperature needed for fusion, the plasma cannot be in contact with solids or liquids, or it would immediately cool. Two general strategies have been developed to isolate and hold a hot plasma and produce fusion energy: (1) Magnetic fusion energy (MFE), discussed in this chapter, where the particles in the ionized gas travel along strong magnetic fields (like beads on a string) and are confined in a magnetic torus and (2) Inertial fusion energy (IFE), where the plasma is rapidly compressed and heated to produce a burst of fusion reactions, confined only by inertia before it flies apart, which is the subject of Chapters 4 and 5.

Magnetic confinement fusion has been pursued as a large-scale international collaboration since it was declassified in 1958, as part of the second "Atoms for Peace" conference. Due to its attractive characteristics and the need for long-term energy supply without greenhouse gas emission, there are large MFE programs in China, Europe, India, Japan, South Korea, Russia, and the US, with smaller programs in other countries. Both Europe and the US developed experiments which produced significant fusion power in the late 1990s, demonstrating that the needed conditions can be achieved. These included producing 10.7 MW peak in a 0.4 sec pulse (TFTR, US)[1] and 16.1 MW peak in a 0.8 sec pulse (JET, EU), where the durations are at half-peak value, with the JET plasma achieving an energy multiplication factor of approximately 0.62.[2] Of course, the input heating power also comes out again, so for example, in the JET experiment, 42 MW of power comes out for 26 MW injected. Heating pulses of 5 sec duration produced 20 MJ of fusion energy. The Japanese experiment JT-60U demonstrated plasma confinement equivalent to an energy multiplier of 0.8[3] but were not able to use tritium fuel. The ITER experiment,[4] under construction, will continue the development of MFE understanding and technology to approximately the scale of a power plant. It is designed to produce up to 500 MW of fusion power with an energy multiplication factor >10, and multiplication factors >5 for pulses of 3000 sec. ITER will prototype the tritium producing blanket modules, the tritium handling system, and other technologies needed for a power plant. The ITER construction and research program is a partnership of the seven large international fusion programs (China, Europe, India, Japan, South Korea, Russia and the US), representing more than half the world's population, and is discussed in Chapter 7.

The available fuel supply for fusion on earth is enormous and very widely distributed. Deuterium atoms make up 156 ppm of all hydrogen atoms on earth. Deuterium is present in all water as HDO molecules, at a level of one per 3,200 light-water molecules. Heavy water is separated in large industrial quantities *via* chemical reactions in several countries as well as by distillation and electrolysis. Lithium is widely distributed in the earth's crust, with reserves at active mines estimated at 13.5 million tons,[5] and total world resources estimated at 39.5 million tons. Lithium is present in seawater at a concentration of approximately 0.2 ppm and a total estimated quantity of 230 billion tons. The South Korean company POSCO has started a program to commercialize extraction of lithium from seawater.[6,7] Even if such a process resulted in lithium production at 20 times the current commercial price, this would add less than 0.1 US cent per kWh to the cost of electricity from fusion. The energy density of lithium in seawater fueling fusion is 2,500 times that of U in seawater-fueling light-water fission reactors. ^{6}Li is the preferred isotope for producing tritium, since it releases energy in the process and can use neutrons of any energy. It has an average natural abundance of ~7.5% of all lithium and has been separated using chemical reactions in several countries. The more abundant ^{7}Li (92.5%) can be retained for other uses, such as batteries. Thus, even if fusion supplies all of the world's electrical energy, the available ^{6}Li should last for millions of years. For both deuterium and lithium, prepared using current technology, the cost of fusion fuel is negligible when compared with the value of its energy content.

In addition to the helium produced by the fusion reaction, radioactive waste is produced due to neutron activation of the material structures near the plasma, including the blanket structure for capturing the neutron's energy and producing tritium from lithium. Studies of fusion power plant conceptual designs have found materials for these components that produce only short-lived radioactivity and could have a radiological hazard potential more than 10,000 times lower than fission-reactor waste[8] if impurities are controlled. Such waste would not require long-term isolation and could qualify for shallow burial disposal.

Fusion is much safer than fission-based energy systems as the amount of fuel available in the plasma is small, typically only sufficient for a few minutes of operation without additional supply injection. Any substantial increase in energy release will cause influx of the wall materials, cooling the plasma and stopping the fusion reactions. Thus no runaway accident is possible. Due to the neutron-activation of the structure surrounding the plasma, design requirements must include measures to prevent loss-of-coolant accidents. However, with proper choice of low-activation materials, the fractional density of radioactive atoms can be low, and a passively safe system can be designed to ensure that radioactive materials cannot be released by after-heat from radioactive decay.

The proliferation risks from magnetic fusion are much less than those from fission.[9] Since no fertile materials, ^{238}U or ^{232}Th, need be near a fusion plasma, it is relatively straightforward for inspectors to detect if the operator attempts to expose such materials to fusion neutrons in order to generate fissile materials that can be used in nuclear explosives. It appears impractical to operate a clandestine fusion system to create such materials, since a large amount of power would be required and the environmental signatures could be easily detected. In the case of a "breakout" scenario, in which safeguard inspectors are expelled, no fissile material should be available at the time of breakout, unlike the case with fission systems. It is possible to disable a fusion power plant remotely, so that it cannot be used to produce fissile material, by damaging power supplies, cooling systems or cryogenic systems, with little risk of dispersing radioactive material. Since the tritium that is produced in fusion systems can be used to boost the yield of nuclear weapons, safeguards for fusion power plants should include not only assurance of the absence of fertile materials but also careful accounting for tritium. There is no overlap between the physics of magnetic fusion and nuclear weapons, so there is no proliferation risk in sharing MFE knowledge and systems.

In magnetically confined fusion power plants, cryogenic superconducting coils will produce the magnetic field to avoid large power-losses from resistive coils. A number of medium- to large-scale fusion experiments around the world have been built using low-temperature superconducting coils and liquid helium cooling, and these will be used for ITER. Liquid helium is also used for cryogenic pumping of the plasma exhaust and it may be used as a thermal-transfer fluid in power plants for high temperature, high efficiency operation. The estimated total available world helium resources are approximately 8.6 million tons,[5] which is much more than that needed for fusion power plants to supply the world's electricity demand. However, since helium is a byproduct of natural gas production, there is a question of whether sufficient helium will be available when fusion energy production is commercialized. This has been analyzed recently[10] using a model of the natural gas market and future production. The authors project that the helium production rate will decrease in the second half of the 21st century. They conclude that there will be sufficient helium to deploy and sustain up to 3.5 TW of fusion power by ~2110, provided that the helium loss-rate is substantially reduced, helium is efficiently recycled on plant shutdown, and methods to substitute for some helium usage are developed. This magnitude of fusion energy could provide the projected total requirement for nuclear energy (fission + fusion) by 2100 to mitigate global warming.[11] The superconducting tokamak in South Korea has already achieved a routine operating helium loss rate of 1%/year (M. Kwon, private communication, 2012), which the study projected as a best-case goal for the year 2100, and they are working to reduce accidental releases. Helium substitution opportunities include the use of high-temperature superconductors with other cryogenic fluids (e.g. Ne), other high-temperature heat-transfer fluids (e.g. CO_2), and mechanical exhaust pumps.

2 MFE Physics and Technology

2.1 *Breakeven, gain and ignition*

Energy can be generated both by splitting the heaviest atoms (fission) and by joining together the lightest ones (fusion). The fuel for almost all fission reactors is the lighter isotope of uranium, ^{235}U. The most favorable fuel for fusion is two isotopes of hydrogen, deuterium (D) (one proton + one neutron), and tritium (T) (one proton + two neutrons). Their reaction is

$$D + T \rightarrow n + {}^4He + 17.6\,\mathrm{MeV}, \tag{1}$$

where $1\,\mathrm{MeV}$ (million electron volts) $= 1.602 \times 10^{-13}\,\mathrm{J}$. About 80% of the reaction energy emerges as kinetic energy of the neutron, while 20% is invested in the helium nucleus, or α particle. Since T does not exist in significant quantities in nature, it is necessary to use the neutrons from fusion to "breed" the required fuel, though the reaction

$$n + {}^6Li \rightarrow T + {}^4He + 4.8\,\mathrm{MeV}. \tag{2}$$

Some neutrons are absorbed in structural materials, and it is necessary to produce slightly more T than is burned in order to provide T to startup future power plants. Thus the primary fusion neutrons are collided with materials such as lead or beryllium, which undergo $(n, 2n)$ reactions: one neutron colliding with a Pb nucleus produces two lower-energy neutrons, each capable of the reaction described in Eq. (2).

The DT fusion reaction rate, in units of reactions per sec per m^3, is given by $n_D n_T \langle \sigma v \rangle_{DT}$, where n_D is the number density of deuterium ($\#/m^3$), n_T is the number density of tritium, and $\langle \sigma v \rangle_{DT}$ is the average over a thermal distribution of particles of the relative particle velocities, multiplied by the cross-section for the DT reaction. Thus the power density in a fusion system is given by

$$P_{\text{fus}}(\text{watts}/m^3) = n_D n_T \langle \sigma v \rangle_{DT} E_{\text{fus}}, \tag{3}$$

where $E_{\text{fus}} = 17.6 \times 1.602 \times 10^{-13}$ J.

Figure 1 shows the fusion power density that can be produced as a function of fuel temperature, for a fixed fuel pressure of 10 atm. Clearly the DT reaction can provide the most power density for a given pressure, but very hot fuel indeed is required to obtain interesting values of power density. At these high temperatures, the fuel is fully ionized: electrons are dissociated from atomic nuclei. The resulting cloud of charged particles is called plasma. Since the positive charge of the fuel nuclei is balanced by electrons, keeping the plasma nearly electrically neutral, the total plasma pressure is given by the sum of the electron and ion pressures.

For a fusion plasma to come to a steady temperature, the input power to the plasma must equal its loss of power:

$$P_{\text{aux}} + 0.2\, P_{\text{fus}} = P_{\text{loss}}, \tag{4}$$

where P indicates total power, not power per m^3, and we have assumed the presence of some "auxiliary" power injected from outside to sustain the plasma, plus that part of the fusion power invested in the electrically charged α particles, which are contained by the magnetic field. The power in the neutrons escapes from the fusion plasma and is absorbed in a "blanket" which surrounds it.

The gain of a steady fusion system is defined as

$$Q = P_{\text{fus}}/P_{\text{aux}}. \tag{5}$$

Magnetic fusion power plants require $Q > 25$ to have a favorable power balance because practical heating systems can have efficiencies in the range of 40%, as can power conversion systems. (As a first approximation we assume that the plasma heating system dominates the power requirements for plant operation.) In this case, four times more fusion power is produced than is required for heating.

$$P_{e,\text{gross}} = 0.4 P_{\text{fus}},$$
$$P_{e,\text{aux}} = P_{\text{aux}}/0.4, \tag{6}$$
$$P_{e,\text{gross}}/P_{e,\text{aux}} = 0.4 \times 0.4 \ P_{\text{fus}}/P_{\text{aux}} = 0.16 \times Q = 4.$$

Ignition in magnetic fusion is defined as $Q = \infty$, which is not required for fusion power plants.

The ITER experiment is designed to bridge the gap in Q from present experiments by operating at $Q = 10$ for pulses of 300–500 sec, at $P_{\text{fus}} = 500\,\text{MW}$, for a total thermal energy production per pulse of $\sim 200,000\,\text{MJ}$. ITER will address not only scientific issues but also many of the technological issues for fusion energy development, as required to handle this high-energy throughput.

2.2 *Magnetic confinement*

At fusion temperatures, the fuel of a fusion system is fully ionized and composed of charged particles, so it responds strongly to the presence of magnetic fields. In particular, both ions (atomic nuclei) and electrons spiral along magnetic fields. Thus a strong toroidal magnetic field can "levitate" an ionized gas, or plasma, and confine its heat. Indeed, experiments around the world have achieved central plasma temperatures of up to 500,000,000°C, well above what is required for fusion (see Fig. 1). Nevertheless, confinement of a fusion-grade plasma entails scientific and technological challenges.[12]

2.2.1 *Transport and turbulence*

The logical primary challenge is to show that it is possible to produce net power from fusion. This means $Q > \sim 6$ [see Eq. (6)], which can be traced back through Eq. (5) to a requirement for low $P_{\text{aux}}/P_{\text{fus}}$, and through Eq. (4) to low $P_{\text{loss}}/P_{\text{fus}}$. However, the core of a fusion-grade plasma must be very hot, as we have seen, and of course the temperature of the material surfaces that surround it must be much cooler, forcing a temperature gradient of order 10^8 °C/m. Strong temperature gradients cause turbulence in fluids, and plasmas are no exception. The turbulence causes particles to travel across magnetic fields lines and carry heat out of the plasma.

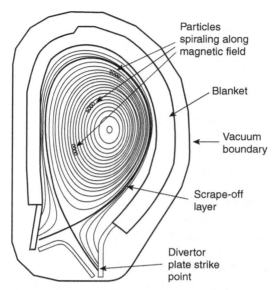

Particles
spiraling along
magnetic field

Blanket

Vacuum
boundary

Scrape-off
layer

Divertor
plate strike
point

Fig. 2.   Cross section at fixed toroidal angle of a tokamak-based magnetic confinement system for fusion energy, identifying the components near the plasma. The nested contours represent the surfaces traced out by the magnetic field lines inside the plasma.

Fortunately, detailed experimental measurements and advanced numerical simulations of plasma turbulence both show that the turbulence intensity varies proportionally to the radius of the particle spirals — the gyro-radius — divided by the radius of the plasma, the ratio of which ratio is denoted as ρ^*. In other words, the tight spirals of the particle motion serve to constrain the turbulence amplitude. The net effect is that the time that energy remains in the plasma increases very rapidly with the size of the plasma and with its magnetic field, indeed as the cube of ρ^*, as shown in Fig. 2.

The "energy confinement time", τ_E^{exp}, is defined in steady conditions as the total energy stored in the plasma divided by the total heating power supplied either through auxiliary heating or through fusion reactions. Modern high-power experiments achieve stored energies of $\sim$10 MJ, with heating powers of $\sim$20 MW, for confinement times above 0.5 sec at 3 T[a] magnetic field. Figure 2 also illustrates that confinement times measured on many

---

[a]Tesla (T) is the unit of magnetic flux density (**B**) in the International System (SI) of units.

fusion systems worldwide form a consistent data set projecting to successful operation of ITER with about 4 sec confinement time at 5.3 T. This performance should provide a wide operational window to explore the physics of high-gain plasmas.

2.2.2 *Stability*

The next scientific challenge is associated with the global stability of the plasma, which is governed by the plasma pressure. For fixed optimum temperature, the fusion power density goes as the square of the fuel density (Eq. (3)) and so, again for fixed temperature, as the square of the pressure. For an economically practical fusion power system, pressures in the range of 10 atm are required. The key parameter governing the physics is denoted as β, the pressure in the plasma divided by the pressure in the magnetic field $(B^2/2\mu_0)$.[b] The magnetic field pressure is limited by magnet technology, so it is desirable to raise β to as high values as practical. ITER, for example, will normally operate at $\beta = 2.5$–3%, while a fusion power plant may require $\sim6\%$, a value already exceeded for short times in current experiments. Experiments in ITER and in devices operated in parallel with ITER will explore this higher range for long pulses.

2.2.3 *Sustainment*

It is desirable to sustain the operation of fusion plasmas for long pulses, in order to minimize cyclic thermal fatigue and mechanical stresses and the need for energy storage. Tokamak plasmas such as ITER require an electrical current to flow around the torus in the long direction, so that the magnetic field lines spiral around the plasma in both the long and short directions, a key requirement for magnetic confinement. This current can be maintained by a combination of transformer action, external injection of power that pushes the electrons relative to the ions, thereby driving a net electrical current, and pressure-driven currents analogous to piezoelectricity. ITER is planned to operate for 300–500 sec without significant external current drive and for up to 3,000 sec with a combination of strong external current drive and pressure-driven current. The challenge of limiting the required current drive power, to maximize Q in this operating mode, will be investigated in ITER and the other superconducting experiments in China, South Korea, Japan, and Germany.

---

[b]μ_0 is the permeability of free space which is equal to $4\pi \times 10^{-7}$ V sec/A m.

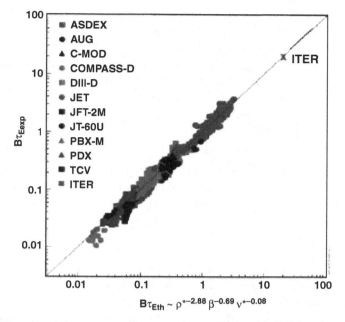

Fig. 3. Experimental energy confinement time, $\tau_E^{\exp}$, multiplied by magnetic field strength, B vs. a multi-experiment confinement time fit expressed in key dimensionless plasma parameters: gyro-radius divided by torus radius, $\rho*$, normalized plasma pressure, β (see text), and collision rate between particles divided by time to circumnavigate the torus, $\nu*$. Projection to ITER is favorable.

2.2.4 *Plasma–material interaction*

With success comes the challenge of handling the 20% of the fusion power that is confined by the magnetic fields, heating the plasma. This heat, along with P_{aux}, flows out of the plasma in a thin "scrape-off-layer" into a divertor region away from the main plasma, ultimately striking divertor plates (Fig. 3).

This configuration has the advantages that it concentrates the He "ash" from the fusion reaction into a region where it can be pumped away, and any particles that are ejected from the material surface are far from the main plasma, reducing contamination of the DT plasma. However, the point of interaction between the thin layer of heat flux and the divertor plate (the divertor "strike point" in Fig. 3) presents a scientific and technological challenge. The leading material candidate for this surface is tungsten, a high-atomic-mass metal. But if the material erodes into the plasma, it can strongly reduce the confinement and reactivity of the system. One alternative is to use a low-atomic-mass liquid metal, such as lithium, at the strike point, which is far less damaging to plasma operation and can be replenished as it erodes. The heat flux from tokamak plasmas is not always steady, but can have high transients. An advantage of a liquid metal surface is that it can withstand a sudden transient heat flux through evaporation and can subsequently be replenished. A melted tungsten surface would likely need to be replaced.

2.2.5 *Neutron–material interaction (including tritium breeding)*

Another consequence of success will be the production of copious amounts of 14.1 MeV neutrons from the fusion reaction. These high-energy neutrons collide with atoms in surrounding materials, displacing them and drive (n, α) reactions in which an energetic neutron causes the production of α-particles, helium nuclei, inside the material. These effects can cause swelling, creep, and embrittlement of structural materials. One can confidently predict based on experiment and calculation that modern low-activation ferritic-martensitic steels developed for fusion applications will operate effectively in the temperature range of 400–600°C, but only at displacements per atom (dpa) up to about 10,[13] which corresponds to about 1 MWyr/m^2 of 14.1 MeV neutrons. Beyond 10 dpa, the effects of the helium produced in the material are uncertain. These steels can be improved through the inclusion of nanosized oxides dispersed at very small scales, with the promise that the nanodispersed particles both strengthen the steel at high temperature and also scavenge He, keeping it from forming bubbles of significant size, and so avoiding embrittlement. Silicon carbide composite materials are also being developed for fusion structures, as they show strong resilience to neutron bombardment.

A DT fusion plasma must be blanketed with material that contains lithium, so that the reaction described in Eq. (2) can proceed. ITER is designed with special ports for "test blanket modules" to study this process. If one takes into account the various ports that are required for plasma heating, and the divertor region that will shield some neutrons from the blanket, it can be challenging to achieve a tritium breeding ratio much in excess of unity, and this is an area of continuing R&D.

2.2.6 *Magnets*

The magnets being constructed for ITER will be the largest and most powerful superconducting magnets in the world. They will produce 5.3 T magnet field in a torus of major radius 6.2 m. The maximum field at the surface of the magnet is designed to be 11.8 T and the stored energy in the magnet is an impressive 41 GJ. This system is based on Ni–Sn supercon-ducting cables surrounded by thick load-bearing conduits, cooled to 4 K. Future fusion magnet systems may strive for higher magnetic field, since fusion power density varies as B^4 at fixed β and plasma temperature. That said, the ITER magnet system is very much in the range of the magnets that will be required for fusion power plants.

2.2.7 *Magnetic field configurations*

Many configurations of the magnetic field have been studied during the fusion program. The highest absolute performance levels have been obtained in tokamaks (Fig. 4) for times of order 1 sec.

Two other configurations that have demonstrated similar performance characteristics are the spherical tokamak and the stellarator. The spheri-cal tokamak is a tokamak with a minor radius that approaches its major radius — a torus with a very small hole in the center. This configuration can run at very high β but its magnetic field is restricted by the small region available for the inner leg of the magnet. It may be a route to lower-cost fusion systems, since the magnets can be relatively inexpensive water-cooled

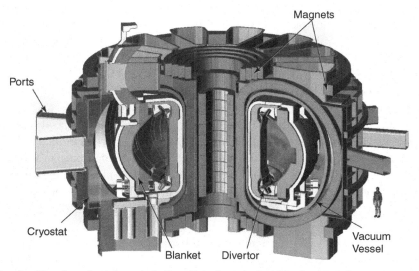

Fig. 4. Drawing of a tokamak fusion pilot-plant design. The horizontal ports are used for external current-drive and heating systems, and for diagnostics. The large top ports are used for maintenance access.

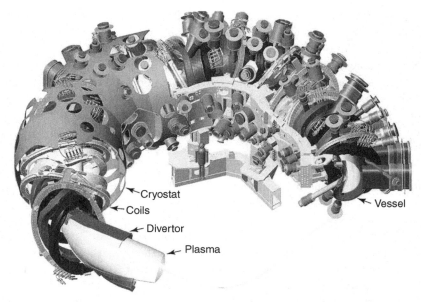

Fig. 5.   Cut-away drawing of the superconducting stellarator experiment, Wendelstein 7-X in Germany.

copper systems and can be joined, allowing easier access for maintenance. However, significant power is required to operate this magnet, reducing Q. Experimental facilities in the US and the UK are investigating the physics properties of the spherical tokamak.

The stellarator is an important complement to the tokamak. In this case the magnetic field lines are coaxed around the torus in the short direction by 3D, non-axisymmetric shaping of the magnets (Fig. 5).

This means that current drive power is not required in stellarators, and it is easier to foresee steady-state high Q operation. Furthermore, stellarator plasmas generally do not create large transient heat fluxes and forces, a significant advantage for the integrity of divertor plates and blankets. Major superconducting stellarator experiments in Japan and Germany will investigate the physics of steady-state, high-β operation in parallel with ITER.

Both the spherical torus and the stellarator configuration are close to the tokamak and share a common theoretical basis, so the improved understanding from ITER results and other experiments should directly apply to all three configurations and inform their use in fusion energy systems.

An ideal magnetic fusion configuration would have such low turbulence that it could be much smaller than ITER and have lower magnetic field; it would support very high β, for high power density at this low magnetic field; it would operate in steady state with little recirculating power; and it would allow exhaust heat to be expelled from the magnet system along diverging magnetic field lines. It would be topologically simply connected, allowing easier access and maintenance. Its performance would be so high that it could burn fuels such as $p-^{11}B$ that generate no neutrons. Such systems would be very challenging for fundamental reasons, but ideas continue to be pursued, and should be, that may have the prospect of exceeding the performance of tokamaks, spherical tokamaks, or stellarators.

3 Progress Toward Fusion Energy

3.1 *National and international research facilities*

The MFE research program is a broad international effort, based on peer-reviewed scientific analysis, with no barriers to communication and replication of results due to classification. Every two years, the International Atomic Energy Agency sponsors a Fusion Energy Conference.[14] In the most recent conference, there were papers presented by experimental and theoretical groups from Australia, Brazil, Canada, China, England, France, Germany, Holland, India, Italy, Japan, Kazakhstan, Russia, South Korea, Spain, Sweden, Switzerland, the US, and Ukraine. The leading research facilities in the world are currently located in China, England, France, Germany, Japan, South Korea and the US (Table 1). Particularly notable are major experiments with superconducting magnetic coils operating in China, Japan, and South Korea, and new large superconducting experiments under construction in Germany and Japan. The largest fusion experiment currently operating is the Joint European Torus, operated by the European Commission in England. This device has about one-half of the linear dimensions of ITER, about two-thirds of the magnetic field strength, and pulse length $\sim$20 sec vs. 300–3,000 sec in ITER.

The progress in energy production in magnetic fusion experiments was exponential from 1970 to 2000, as shown in Fig. 6. Since that time, no MFE experiments have operated with DT plasmas, as the world awaits the completion of ITER, as discussed in the next chapter.

Table 1: Facility Parameters for the Major MFE Experiments, Worldwide, Including Experiments Under Construction and Being Upgraded.

Location	Type	Major radius (m)	Minor radius (m)	Magnetic field (T)	Max. heating power (MW)	Max. pulse length (sec)	Comments
China	Tokamak	1.85	0.45	3	26	1,000	Superconducting
England	Tokamak	3.1	1	4	51	20	EU project DT capable
England	Spherical Tokamak	0.85	0.65	0.84	7.5	4	Stage 1 of upgrade project
France	Tokamak	2.5	0.5	3.7	17	1,000	Superconducting
Germany	Tokamak	1.65	0.5	3.3	34	10	Upgrading
Germany	Stellarator	5.5	0.5	3	20	1,800	Superconducting
Japan	Stellarator	3.7	0.6	3	36	3,600	Superconducting
Japan	Tokamak	2.97	1.2	2.25	39	100	Superconducting Under construction
Russia	Tokamak	1.48	0.67	2	23	10	Under construction
South Korea	Tokamak	1.8	0.5	3.5	24	300	Superconducting
US	Tokamak	0.68	0.22	5.4	8.5	5	$B = 8T$ at reduced pulse length
US	Tokamak	1.67	0.67	2.2	35	10	Includes proposed power and pulse length upgrades
US	Spherical Tokamak	0.9	0.6	1	16	5	$P = 21$ MW for shorter pulse
ITER	Tokamak	6.2	2	5.3	75	300–3000	Superconducting Under construction

Note: The pulse lengths and heating powers listed are the maximum expected after currently planned upgrades.

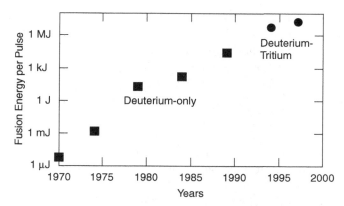

Fig. 6.   Energy produced per pulse in magnetic fusion experiments worldwide, 1970–2000.

3.2 *Theory and modeling*

It is important to recognize that advances in fusion energy production have been accompanied by similar advances in theoretical understanding and computational modeling of magnetic fusion plasmas. These advances were required to extrapolate with confidence from present experimental results to ITER. Theoretical codes now predict the observed properties of turbulence in the hot core of tokamak plasmas, including the absolute rate at which energy is lost. Stability limits are routinely calculated with high precision, and the current drive techniques required for sustainment are understood to high accuracy. The physics of the plasma edge and scrape-off region may be the most mysterious, but recently, theoretical models have been able to capture both the magnitude and scaling of the scrape-off width for the first time. Advanced computation has been critical for the design of new materials to operate in the 14 MeV neutron environment, and for interpreting experimental results from neutron exposure. And computational tools are used extensively in the design and optimization of powerful superconducting magnet systems.

4 Development Plans and Design Studies

4.1 *National and international development of fusion energy*

ITER will substantially advance our knowledge of burning plasmas and technical capabilities toward practical fusion energy. However, ITER is designed as an experiment, not a prototype power plant. With the launch of ITER construction and preparations for its operation, several of the ITER partner countries and country-groups have established programs to resolve the remaining challenges, and to design the experiments and facilities needed to advance to a demonstration magnetic-fusion power plant (Demo). The remaining challenges have been identified and extensively documented (e.g. see Ref. 13), and fall into three broad groups:

- *Reliable, predictable, steady-state operation at high performance.* In order to produce substantial net electricity, the fusion power multiplier must be much higher than the steady-state value of 5 expected for ITER. This requires that the plasma pressure (β) be higher for Demo. It also requires that less external current drive power be used, and that more current be self-generated by the plasma pressure, which requires increased β. Higher β increases instabilities, is more challenging to control, and has not been demonstrated in tokamaks with steady-state current operation. This is a research focus for superconducting tokamaks worldwide.

  In principle, this is not a challenge for stellarators, which do not need external current drive and have demonstrated steady high-β operation. However, stellarators have not demonstrated adequate confinement simultaneously with high-β. This is a research focus for the superconducting stellarator in Japan and for the new superconducting stellarator in Germany.

- *Reliable exhaust of plasma heat and helium ash.* The steady heat flux in the exhaust structures will be much higher for Demo than for ITER. In addition, fusion neutron bombardment and transmutation will degrade the properties of solid materials, including their heat conductivity. Plasma instabilities can generate large transient increases in the exhaust flux. Either the instabilities must be eliminated or the transient fluxes absorbed. The high heat flux in Demo must be handled without excessive erosion or infiltration of the plasma by the wall material. Research directions include developing new materials, methods to spread the edge magnetic field to spread the exhaust heat flux, methods to cool the edge plasma to reduce erosion, and possible use of liquid surfaces to self-heal damage and handle higher power fluxes. Development of these strategies may require dedicated experiments.

- *Capturing neutrons and self-sufficient production of tritium.* The energy from the fusion reactions must be captured and converted to electricity, including both the 14.1 MeV neutrons and the α particles (helium nuclei). The neutrons must also interact with lithium in the blanket to produce tritium. The tritium must be gathered, purified, stored, and injected into the plasma for continued fusion burning. The 14.1 MeV neutrons bombard the structural materials around the plasma and blanket, causing dislocation damage and transmuting the constituent atoms, including generating internal gas bubbles. Materials and designs must be validated as being able to survive and function safely in this environment. In addition, the blankets must shield the superconducting coils from the neutrons, to prevent conductor and insulator damage. These issues will be studied on ITER, including its Test-Blanket Module (TBM) program. Studies with higher neutron fluence will need dedicated neutron sources, such as the proposed International Fusion Materials Irradiation Facility (IFMIF), and may require staged studies on facilities approaching Demo.

The solutions to these challenges must be integrated and made reliable and maintainable with high availability to have a practical, economical energy system.

Possible solutions to these integration challenges have been explored in many pre-conceptual design studies for possible fusion power plants, based on tokamaks, stellarators and other magnetic configurations. Recent examples include the EU PPCS studies,[15] the US ARIES ACT studies,[16,17] and the Japanese SlimCS study,[18] which explore different design choices and development risks.

Several national and international programs have been launched to address the post-ITER challenges and prepare for a Demo power plant, expected to be the last step before commercialization. Differing needs for new energy sources drive the various national programs, producing differences in plan details and designs. Nations with the most time-urgent needs are planning larger steps and focused programs. However, all of the programs aim to provide fusion power by approximately mid-21st century. The plans and development of the various Demo programs are reviewed and compared in periodic workshop organized by the IAEA.[19] Pre-conceptual designs for major facilities after ITER are summarized in Table 2. The major Demo-related programs include:

- The Broader Approach[20] is a joint program by the EU and Japan to carry out R&D to prepare for Demo and complement ITER, established by signed agreement in 2007. The Broader Approach has three major parts: (1) a new superconducting long-pulse tokamak, JT-60SA, to explore steady-state high-β operation and optimization of the plasma configuration for Demo. (2) Design, R&D, and prototype development for

Table 2: Facility Parameters ITER and Preconceptual Designs for Burning-plasma Facilities after ITER, towards Demo. It is Likely that the Parameters will Evolve as R&D Continues.

		Major radius (m)	Minor radius (m)	Magnetic field (T)	Fusion power (MW)	Electric power (MW)	Comments
	ITER	6.2	2	5.3	500	0	pulsed, <3000 sec
EU	EU-Demo	9	2.25–3	7.4–5.3	1800	500	pulsed, 0.6–1.5 hours
China	CFETR-I	5.7	1.6	4.5–5	~200	0	continuous
	CFETR-II	6	2	6	1200	~300	continuous, Demo validation
Korea	KDemo-I	6.8	2.1	7.4		~150	continuous, Demo validation
	KDemo-II	6.8	2.1	7.4	2200–3000	~500	continuous

IFMIF, the proposed accelerator-based intense 14.1 MeV neutron source for testing materials and sub-components for future fusion power systems. (3) The International Fusion Energy Research Center, which includes Demo design studies, Demo R&D coordination, a shared supercomputer center, and remote operation facilities for ITER.

• The Euro-Fusion Program[21] is coordinating the R&D activities in the EU-member states preparing for ITER research and for Demo design. This effort is part of the European Horizon 2020 Program, and has the goal of demonstrating fusion generated power entering the electricity grid before 2050. A roadmap of the required R&D was developed and documented,[22] and is organized into eight strategic missions with milestones:

(1) Plasma regimes of operation
(2) Heat exhaust systems
(3) Neutron resistant materials
(4) Tritium self-sufficiency
(5) Implementation of the intrinsic safety features of fusion
(6) Integrated DEMO design and system development
(7) Competitive cost of electricity
(8) Stellarator

The top-level milestones include starting Demo construction in 2030 and operation in ~2041. A pre-conceptual design for Demo has been developed,[23] based on the ITER approach needed to design a Demo, and be prepared to start construction in approximately 2030. The program includes a comprehensive program of technology development, using additional smaller-scale facilities.

- China completed a multi-institutional preconceptual study of the Chinese Fusion Engineering Test Reactor (CFETR),[24] as requested by their Ministry of Science and Technology. This facility would operate in two phases, separated by a substantial upgrade. The goal of the first phase is to operate at high plasma performance in steady-state or long-pulse with high-duty factor, demonstrate self-sustaining tritium breeding, and demonstrate durable first-wall and blanket materials at high heat load and neutron flux. The second phase would validate the needed capabilities and validate the integrated design for Demo with Q>10 and ~1 GW of fusion power. A road-map for fusion development in China has been developed[25] that will start construction of CFETR in 2020, start operation in 2030, and upgrade it for phase-II in 2040. A prototype power plan would be built in 2050–2060.
- South Korea has identified the R&D activities needed to design a Demo fusion reactor,[26] and started the pre-conceptual design of K-Demo.[27] This facility would have two operating phases separated by a substantial upgrade. The first phase would operate in steady-state at high performance, demonstrate self-sustaining tritium breeding, test system components, and generate a modest amount of net electricity. The second phase would be full Demo operation, producing ~500 MW of net electricity, and demonstrate the characteristics needed for competitive cost of electricity, which would Demo preconceptual design study with a goal to begin construction in the mid-2020s and operation in the mid-2030s.
- Japan recently completed a multi-institutional study[28,29] of the needed R&D to establish the technical basis for developing a fusion Demo reactor, and the strategy needed to conduct it. The study was requested by the Japanese ministries. The decision to start Demo construction will await ITER achieving high performance milestone (Q>20), sustained moderate performance (Q>5), and other technical achievements, and is currently expected to occur in ~2030. At that time, the magnetic configuration will also be decided, based on conceptual Demo design studies and their technical basis. The next phase of the Japanese program planning will be the development of a fusion development roadmap, which will be undertaken by a government committee.

Other countries, including India, Russia, and the US, are evaluating facilities to conduct specific research to prepare for an eventual Demo. These include neutron sources, integrated test facilities, and upgrading of existing facilities. In addition, individual laboratories and collaborations are engaged in design studies and R&D program planning.

4.2 *Private development of fusion energy*

A number of private companies are engaged in research toward achieving fusion energy with significant private funding. Examples include General Fusion,[30] Lockheed Martin,[31] Tokamak Energy,[32] Tri-Alpha[33] and others. The increase in these activities is due to the need for new energy sources, particularly ones that do not contribute to climate change. The private companies, generally, are examining ways to simplify the engineering design from the beginning, in order to develop less-expensive, more maintainable systems. Applying the improved understanding of plasma physics and plasma modeling, they are searching for ways to improve hot-plasma confinement in a simple system to the level needed for fusion. Several strategies are being used:

- Compact magnetic configurations, to minimize size and cost,
- Eliminate or minimize the toroidal component of the magnetic field,
- Pulsed magnetic compression, on approximately a millisecond timescale.

While these approaches have not yet achieved the temperature and pressure needed for fusion gain, they have produced novel innovations, improved understanding of magnetic confinement configurations and new fusion technological approaches.

5 Summary

Controlled fusion offers an effectively inexhaustible energy source without production of greenhouse gases or long-lived radioactive waste, without the possibility of catastrophic runaway accidents, and with low proliferation risk. It can be made available to all nations. There has been dramatic progress in understanding the physics and technology needed for MFE, and producing the conditions needed for substantial fusion power production. The next major step will be ITER, providing industrial levels of fusion power for long pulses, and enabling the study of burning plasmas and fusion power technology. Beyond ITER, the challenge will be to make fusion practical and economical. Several countries are already actively engaged in R&D

and design programs for post-ITER fusion power systems. A number of private companies have started to explore ways to possibly produce fusion energy using simpler engineering approaches. These activities are taking the steps needed to achieve fusion's potential as an attractive new energy source.

References

1. R. J. Hawryluk (1998). Results from deuterium-tritium tokamak confinement experiments. *Rev. Mod. Phys.*, **70**, 537.
2. J. Jacquinot (1999). Deuterium-tritium operation in magnetic confinement experiments: results and underlying physics. *Plasma Phys. Control Fusion*, **41**, A13.
3. T. Fujita and the JT-60 Team (2003). Overview of JT-60U results leading to high integrated performance in reactor-relevant regimes. *Nucl. Fusion*, **43**, 1527.
4. The ITER Organization. Accessed on August 24, 2015 at http://www.iter.org/.
5. USGS Commodity Summaries (2015). Accessed on August 24, 2015 at http://minerals.usgs. gov/minerals/pubs/commodity/.
6. POSCO news release, 5 February (2010). Accessed on September 19, 2012 at http://www.posco.com/homepage/docs/eng/jsp/prcenter/news/s91c1010025v.jsp?mode=view &idx=1272.
7. *The Korea Times*, 23 February (2012). Accessed on August 24, 2015 at http://www.koreatimes.co.kr/www/news/nation/2012/02/182_105554.html.
8. M. Kikuchi and N. Innoue (2001). Role of fusion energy for the 21 century energy market and development strategy with International Thermonuclear Experimental Reactor" in *WEC-18*, World Energy Council.
9. A. Glaser and R. J. Goldston (2012). *Nucl. Fusion*, **52**, 043004.
10. R. H. Clark and Z. Cai (2012). in *The Future of Helium as a Natural Resource*. eds. W. H. Nuttall, R. H. Clarke, and B. A. Glowacki. Routledge, London and New York, p. 235.
11. R. J. Goldston (2011). *Science & Global Security*, **19**, 130.
12. US DOE, Research needs for magnetic fusion energy sciences (2009). Accessed on August 24, 2015 at http://science.energy.gov/~/media/fes/pdf/workshop-reports/Res_needs_mag_fusion_report_june_2009.pdf.
13. Fusion Energy Sciences Advisory Committee Report on Opportunities for Fusion Materials Science and Technology Research Now and During the ITER Era, February (2012), p. 68. Accessed on August 24, 2015 at http://science.energy.gov/_/media/fes/pdf/workshop-reports/20120309/FESAC-Materials-Science-final-report. pdf.
14. Accessed on August 24, 2015 at http://www-pub.iaea.org/iaea meetings/46091/25th-Fusion-En- ergy-Conference-FEC-2014.
15. D. Maisonnier *et al.* (2007). *Nucl. Fusion*, **47**, 1524.
16. F. Najmabadi *et al.* (2006). *Fusion Eng. Design*, **80**, 3.

17. F. Najmabadi and A. R. Rafray (2008). *Fusion Sci. Technol.*, **54**, 655.
18. K. Tobita *et al.* (2009). *Nucl. Fusion*, **49**, 075029.
19. G. H. Neilson, G. Federici, J. Li, D. Maisonnier, and R. Wolf. *Nucl. Fusion*, **52** (2012), 047001. Accessed on August 24, 2015 at http://advprojects.pppl.gov/ROADMAPPING/presentations.asp.
20. Euratom and Japan. Broader Approach Activities in the Field of Fusion Energy Research. Accessed on August 24, 2015 at http://www.ba-fusion.org/.
21. EUROfusion Consortium. European Consortium for the Development of Fusion Energy. Accessed on August 24, 2015 at https://www.euro-fusion.org/.
22. F. Romanelli (2012). Fusion electricity — a roadmap to the realization of fusion energy. EFDA, Accessed on August 24, 2015 at https://www.euro-fusion.org/wpcms/ wp-content/uploads/2013/01/JG12.356-web.pdf
23. G. Federici *et al.* (2014). *Fusion Engr. Design*, **89**, 882.
24. Y. Wan *et al.* (2014), IEEE *Trans. Plasma Sci.* **42**, 495.
25. Y. Wan, presentation at Fusion Power Associates, 16 Dec. 2014. Accessed on September 21, 2015 at http://fire.pppl.gov/FPA14_Chinese_CFETR_Wan.pdf.
26. G.S. Lee, presentation at Fusion Power Associates, 16 Dec. 2014. Accessed on September 21, 2015 at http://fire.pppl.gov/FPA14_Korean_Fusion_Prog_GSLee.pdf.
27. K. Kim *et al* (2015). *Nucl. Fusion*, **55**, 053027.
28. Joint-Core Team for the Establishment of Technology Bases Required for the Development of a Fusion DEMO Reactor. Basic Concept of DEMO and Structure of Technological Issues. 19 Jan. 2015. Accessed on September 21, 2015 at http://www. naka.jaea.go.jp/english/report/201505_1st.pdf.
29. H. Yamada, A. Ozaki, R. Kasada, R. Sakamoto, Y. Sakamoto, H. Takenaga, T. Tanaka, H. Tanigawa, K. Okano, K. Tobita, O. Kaneko, and K.Ushigusa. Joint-Core Team for the Establishment of Technology Bases Required for the Development of a Fusion DEMO Reactor. Chart of Establishment of Technology Bases for DEMO. 1 Mar. 2015. Accessed on January 29, 2016 at http://www.naka.jaea.go.jp/english/report/201501report_2.pdf.
30. General Fusion, Inc. Accessed on September 21, 2015 at http://www.general fusion.com/.
31. Lockheed Martin Corp. Accessed on September 21, 2015 at http://www.lockheedmartin.com/us/products/compact-fusion. html.
32. Tokamak Energy Ltd. Accessed on September 21, 2015 at http://www.tokamakenergy.co.uk/.
33. M. W. Binderbauer *et al.* (2015). *Phys. Plasmas*, **22**, 056110.

Chapter 7

Creating a Star — The Global ITER Partnership

M. Uhran

US ITER Project Office
Oak Ridge National Laboratory
Oak Ridge, TN 37831, USA
uhranml@ornl.gov

International thermonuclear experimental reactor (ITER) is an unprecedented global partnership to demonstrate the scientific and technological feasibility of generating, carbon-free, and virtually unlimited energy through the fusion of hydrogen isotopes. Now under construction in southern France, the ITER fusion reactor is designed to achieve and sustain self-heated, or "burning", plasma that can produce 10 times more power than required for plasma heating. This overview provides an introduction to fusion, summarizes the history of the ITER project, describes key subsystems and elements of the ITER reactor, and includes objectives and goals of the ITER research plan.

1 Introduction

Bolts of lightning arcing down across the horizon of the African savannah were as mysterious to early humans, as the Sun during the day, and the veil of stars across the blackness of the night sky. The fact that these phenomena are closely related as an energetic state of matter — now known as plasma — would take many more millennia before being scientifically understood. Nevertheless, human recognition that a bolt of lightning striking a lone tree produced a useful fire signaled the dawn of civilization's endless pursuit of energy resources.

Wood served as an energetic fuel of human society for at least 10 millennia and then, in the late 19th century, human understanding of the

nature of fuels began changing rapidly. Coal, the petrified remains of biode-
graded organic matter, fed the powerhouses that drove a global industrial
revolution. By the early 20th century, petroleum derivatives began joining
coal as a more transportable and efficient fuel, mobilizing the accouterments
of war, and empowering the nations that mastered oil and gas extraction
as world leaders.

In 1905, Albert Einstein revolutionized the world of physics with publication of the mass–energy equivalence principle, $E = mc^2$, and by the early
1950s, the first nuclear power plants began making their debut. Nuclear
energy based on uranium fission — generating low-cost electricity (i.e.
5–10% per kilowatt-hour) over long plant life-cycles (i.e. 40–60 years) at
unprecedented power levels (i.e. 2–3 gigawatts per plant) — now represents
a major carbon-free energy source in the modern world. However, the evolution of energetic fuels does not necessarily have to end at the close of the
20th century with the industrialization of uranium fission. Hydrogen fusion
holds the potential to be the next major development in nuclear energy.

Uranium is a "heavy element" (atomic number: 92) that releases energy
when the nucleus is split to form smaller nuclei. The energy is released
primarily in the form of neutrons, gamma rays and energetic fragments. In
contrast, hydrogen is a "light element" (atomic number: 1), that releases
energy when the nuclei of hydrogen isotopes are fused together to form
helium (atomic number: 2). A technological transition from heavy-element
fission to light-element fusion would lead to not only far fewer radioactive
isotope byproducts, but also increased global availability of natural fuel.
Inputs would come in the form of "heavy" hydrogen — termed deuterium
because it has one additional neutron — which is found in seawater, and
lithium, which can be extracted from the Earth's crust and oceans. These
two elements are the feedstock for a fusion energy fuel cycle.

Achieving controlled hydrogen fusion in order to generate carbon-free
electricity is one of the "grand challenges for engineering in the 21st century" according the U.S. National Academy of Engineering.[1] As with many
historic advances, the emergence of several disruptive technologies can often
converge and precede a practical solution. In the case of fusion, the close
of the last century brought with it the availability of high performance
computing that enabled great strides forward in the modeling of complex
large-scale flows, turbulent small-scale flows, and energetic particle dynamics, all of which bear acutely on fusion processes. In combination with highly
sensitive diagnostic instruments, theoretical models can be formulated, validated and integrated far more effectively than in the past. Lastly, high-
efficiency, low-temperature superconductors, such as niobium-tin (Nb_3Sn),

have replaced copper in the electromagnetic coils used for confinement and shaping of fusion plasmas. As such new technologies are applied to the engineering of 21st century reactor configurations, fusion need no longer be "thirty years away".

In the sociopolitical context, hydrogen fusion offers the benefits of nuclear energy without the constraints associated with fission. The primary by-product is inert helium, while the tritium that is also produced by neutron bombardment of lithium remains contained in a closed fuel cycle. Therefore, fusion can be a very safe, carbon-free source of electricity. Fusion also requires high temperatures; any disturbance tends to degrade confinement, cool the plasma and reduce the reaction rate, so there is no risk of a "runaway reactor" or "meltdown". Nor is there any highly radioactive fuel that keeps emitting heat after shutdown. Thus, fusion reactors would be inherently safer to operate than fission reactors. While the highly energetic flux of 14.1 MeV fast neutrons will activate plant structure and materials, the radioactive lifetime can be engineered through low-activation materials selection, thus reducing radiotoxic lifetimes to readily managed durations (e.g. a century as opposed to millennia).[2] Finally, hydrogen is a plentiful element making up an estimated 90% of all atoms and three quarters of the mass of the visible universe; it is the third most abundant element on the Earth's surface (behind oxygen and silicon). A hydrogen fusion-based energy source would be globally accessible without the restrictions of geopolitically-controlled natural resources.

2 The ITER Partnership

International Thermonuclear Experimental Reactor (ITER) — Latin for "the way" or "the journey" — can trace its origins to the wake of the 1970s global energy crisis. In October 1973, the Organization of Petroleum Exporting Countries announced a trade embargo on oil that jolted the world at large into sudden recognition of energy as a fundamental factor of national production and security. This was followed by a second "oil shock" in 1979 associated with decreased oil production as a consequence of the Iranian Revolution. Prior to that decade, energy resources had rarely been considered as a constraining factor on national economies.[3]

2.1 *From INTOR to ITER*[4]

In 1978, the International Atomic Energy Agency (IAEA) invited member governments involved in fusion research to consider the timeline for fusion energy development and potential advantages of international cooperation on fusion science and engineering. In response, Evgeny Velikhov of the former Soviet Union proposed an international cooperative project to design, construct and operate an experimental reactor based on the tokamak concept. The International Fusion Research Council, an advisory body to the IAEA, recommended a Specialist Committee be formed, including representatives from the U.S., Europe, Russia, and Japan, to evaluate the prospects. At an organizational meeting of the committee in Vienna, Austria, the plan for a series of workshops was agreed upon and the effort was christened the International Group Working on a Tokamak Reactor (INTOR).

During the decade from November 1978 to March 1988, many sessions of the INTOR Workshops were conducted, with contributions from as many as 150 individuals from the participating parties. By 1985, however, plasma physicists and fusion engineers had become concerned that the INTOR Workshops might amount to little more than a "paper study." This changed during the 1985 Geneva Summit Meeting, when USSR Head-of-State Mikhail Gorbachev proposed to U.S. President Ronald Reagan that a joint effort be undertaken to advance the INTOR concept into a final design, followed by construction and operations phases. Within a year, the ITER Project was conceived and there was an intergovernmental agreement to proceed. Meanwhile, the INTOR Workshops continued and culminated in publication of a final report in 1988.[5]

The ITER Project officially commenced in 1988 and was initially hosted at the Max Planck Institute for Plasma Physics near Munich, Germany for the early Conceptual Design Activity (1988–1991). An ITER Engineering Design Activity (EDA) period followed (1992–1998) where the US., Europe, Russia, and Japan formalized their national participation, and many of the scientists who had participated in the INTOR Workshops continued as members of the new ITER global team.

The US did not extend participation beyond the 1998 end of the EDA. Subsequently, a series of US-sponsored workshops over 2001–2002 culminated in a 2-week study session in Snowmass, Colorado in Summer 2002, for the purpose of seeking expert assessments of the scientific and technological readiness for studying burning plasmas and of three approaches to that study, including the ITER version that emerged from the EDA. The ITER approach was selected by the U.S. government based on collective judgment of participating experts, including the Fusion Energy Sciences Advisory Committee to the U.S. Department of Energy, as well as a study completed by the U.S. National Academies/National Research Council.[6] Consequently, in 2003 the U.S. rejoined ITER negotiations. China and

South Korea also joined in 2003, and India completed the current partnership by joining in 2005.

2.2 *Provisions of the joint implementation agreement (JIA)*[7]

Under the auspices of the IAEA, a joint implementation agreement was formally signed among EURATOM (the European Atomic Energy Community), the Republic of India, Japan, the People's Republic of China, the Republic of Korea, the Russian Federation and the United States of America in November 2006. According to the provisions, an International Fusion Energy Organization (IO) was established for the purpose of demonstrating *"the scientific and technological feasibility of fusion energy for peaceful purposes, an essential feature of which would be achieving sustained fusion power generation"*. Following years of study and negotiation among the parties, the JIA also resolved that the IO and tokamak research and development (R&D) laboratory be situated at St. Paul-lez-Durance in the south of France, adjacent to the Commissariat à l'énergie atomique (CEA), the French Atomic Energy Commission.

The IO is governed by a Council composed of up to four representatives from each of the seven partners. Each government partner has also designated a "Domestic Agency" (DA), in order to provide contributions to the IO through an established legal entity. Contributions are defined in two forms: (a) in-kind goods and services, consisting of specific components, equipment, materials, and R&D, as assigned to each partner for delivery in accordance with IO technical specifications, and (b) in-cash contributions to cover the IO annual operating expense and components to be supplied directly by the IO. Annexes to the JIA provide further detail on specific technologies to be contributed, etc.

The JIA was approved with an initial duration of 35 years and included a provision for a Special Committee to be formed eight years prior to expiration, in order to advise on extension in light of progress achieved.

3 Project Life Cycle

Based on agreement among the international parties, the project life cycle is divided into three distinct phases:

Phase 1: Design and Construction.

Phase 2: Operations.

Phase 3: Decommissioning.

3.1 *Design and construction phase*

The point-of-departure for this first phase was the configuration resulting from the 1992–2001 EDA period. The final design and construction phase includes:

- Establishment and operation of the IO as an institutional entity;
- Resolution of details on functional and design specifications;
- System engineering and analysis;
- Completion of procurement arrangements between the IO and DAs;
- Site survey, regulatory approvals, excavation and development;
- Building and road planning, civil engineering and construction;
- Prototype development, testing and evaluation for plant subsystems and elements;
- Manufacturing and delivery of plant subsystems and elements to construction site;
- Subsystems and elements integration and assembly of plant.

3.2 *Operations phase*

The primary mission of ITER is experimental research operations; details for the operations phase are therefore discussed in a later section on the Research Plan. The operations phase commences following completion of construction and gradually progresses from commissioning of systems to full safety-qualified nuclear R&D operations as the nuclei fuel mix changes from hydrogen and helium to deuterium and tritium. During the operations phase, the IO is responsible for establishing a fund to provide for decommissioning of ITER facilities.

3.3 *Decommissioning phase*

The decommissioning fund and facilities are to be transferred to France, as the host state, following completion of research operations. France will remain bound to Article 20 of the JIA, which constrains any further uses of ITER facilities and equipment to peaceful purposes, and ensures conformity with principles of non-proliferation.

4 Fusion R&D laboratory complex, St. Paul-lez-Durance, France

The ITER complex resides on a 180-hectare tract of land, including an elevated 42-hectare platform standing 315 m above sea level, located approximately 60 km northeast of the Port of Marseille in St. Paul-lez-Durance, France. The site was selected for its proximity to the existing Cadarache CEA facility, which has been a French scientific research center for nuclear energy since 1959.

Two years of site preparations were completed in June 2009, and by May 2010 an architect engineering contract was awarded for design and construction of buildings, infrastructure and power supplies to the European consortium ENGAGE (Assystem, France; Atkins, UK; Empresados Agrupados, Spain; and Iosis, France). Site development commenced with the building of roads, excavation for the first foundations of the tokamak complex, and construction of a power substation.

The ITER Site Master Plan includes 39 buildings and a wide range of supporting infrastructure necessary to support research, operations, and maintenance of the ITER tokamak facility.

5 Status of the ITER Project (2015)

On November 9, 2012, the French Ministry of Ecology, Sustainable Development and Energy issued a decree authorizing construction of the ITER facility; this document granted the IO a license to construct a nuclear facility. During the period 2013–2015, the worksite was transformed from a sparse construction platform into a busy hub of industrial activity with foundations being poured, steel exoskeletons erected, and sky cranes rising above the future tokamak pit. The Poloidal Field (PF) Coils fabrication building and a Cryostat Workshop, where the approximately 30 × 30 m stainless steel structure will be assembled by India, were completed, while the Assembly Hall, Tokamak and Diagnostics Buildings are rising up from the ground. Figure 1 provides an aerial view of the ITER construction site taken in August 2015 and Fig. 2 provides an artistic rendition of the ITER complex upon completion.

In parallel with the ramp-up in site construction activity, DAs from around the world began fabrication of early-lead components. Toroidal field (TF) cable-in-conduit superconductor entered production in China, Europe, Japan, Korea, Russia, and the U.S., and shipments to coil winding facilities began. Figure 3 shows a winding of TF conductor into a D-shaped coil at the European winding facility in La Spezia, Italy. Over 80,000 km of superconducting strand (Nb_3Sn) will eventually be used to wind the 18 TF coils that form the electromagnetic field surrounding the torus and confining the plasma. In the U.S., the central solenoid (CS) coil fabrication facility was completed and the first of seven modules entered production.

Fig. 1.   Aerial view of the ITER construction site — April 2015.
Source: ITER organization.

Fig. 2.   Artists' rendering of the ITER complex upon completion.
Source: ITER organization.

Fig. 3.   Winding of TF superconducting cable-in-conduit into D-shaped coils at the European winding facility in Spezia, Italy.
Source: F4E.

Fig. 4.   Superconducting cable-in-conduit spools and tooling stations for winding the central solenoid at the general atomics magnet technologies center in Poway, CA.
Source: GA.

Figure 4 is a photo of the new General Atomics Magnet Technologies Center where the 18-m high, 4-m diameter, 1,000 ton superconducting magnet will be produced. The CS is figuratively the "heartbeat of ITER", because the pulse it generates drives the current in the plasma. In shipping, the first U.S. procured and fabricated "highly exceptional loads" (over-sized loads) also began arriving at the ITER site, including 87-ton high-voltage transformers required by the steady-state electrical network and 61,000-gallon drain tanks for the tokamak cooling water system.

By 2015, the ITER Project was on the cusp of transition from design to construction with preparations well underway to begin assembling the tokamak. Construction process documents and work packages included stamps of approval, and a global workforce was engaged across the partnership. The IO was in the process of transforming its focus from design to construction management as the assembly phase approached.

6 Tokamak Design

The term "tokamak" originated at the Kurchatov Institute in Moscow as an acronym for "toroidal chamber with magnetic coils". It was based on a theory of electromagnetic traps proposed by Oleg Lavrentiev while attending the Kharkiv Theoretical Physics School in the Ukraine. Soviet physicists Igor Tamm and Andrei Zhakarov, under the direction of Lev Artsimovich, led early advancement of theory to practice.[8]

An early challenge to sustaining conditions for fusion research was reducing the heat loss by confining the hot (i.e. 150–300 million°C) plasma for sufficient time at necessary density. The use of an electromagnetic field shaped as a torus proved to be effective in overcoming this constraint. As a result, the tokamak evolved to become the most practical and well-understood experimental device for fusion R&D. Approximately 200 tokamaks have been constructed around the world with more than thirty remaining in operation today.[9]

In order to sustain the generation of power and deliver useful energy, a practical fusion process must yield more energy than required to start and maintain the reaction. Attaining the required "Q-value" (ratio of fusion power produced to power required) is therefore dependent on reducing the external heating power by achieving a self-heated, or "burning", plasma state, where the energy released from fusing hydrogen nuclei is sufficient to dominate the heating needed to sustain the reaction. Although power production through controlled fusion reactions was experimentally proven for brief instants during the 1990s,[10] generation of fusion power by sustaining burning plasma remains to be demonstrated.

For these reasons, achieving and maintaining substantially self-heated, or "burning", fusion plasma is the mission objective of ITER. Realization of the objective would represent an historic turning point in fusion energy R&D. This most salient aspect was clearly articulated by the U.S. National Research Council in 2004: "*It is widely agreed in the plasma physics community that the next large-scale step in the effort to produce fusion energy is to create a burning plasma — one in which alpha particles from the fusion reactions provide the dominant heating of the plasma necessary to sustain the fusion reaction*".[11]

ITER is designed to be the first tokamak to produce burning plasma. Since achieving that state demands a large plasma volume, ITER will also be the largest tokamak yet. The rationale for the greatly enlarged scale (i.e. more than eight times the plasma volume of prior tokamaks) was based on the advantage gained by reducing the ratio of toroidal surface area to volume in order to minimize heat loss. An earlier ITER tokamak design, of yet larger toroidal geometry, was reduced in scope due to a trade between cost and scale. The current ITER design is most notable for its flexibility in supporting a wide range of experimental operations. The arrangement of toroidal, poloidal, correction, and in-vessel coils permits unprecedented reactor-scale spatial control over electromagnetic field stability, drift, and disturbances, while the CS coil provides variable current pulsing and plasma shaping. Multiple heating subsystems allow tuning of the depth of electron- and ion-heating, and driving of plasma current non-inductively throughout the toroidal cross-section. An exhaustive array of diagnostic instruments and associated actuators will provide active control techniques with which to experimentally influence the plasma shape, thermal and density gradients, current pulse and other key parameters. The integrated system represents a powerful tool for fusion R&D. Major physical characteristics and functional performance parameters are summarized in Fig. 5.

A tokamak configuration will not, *a priori* become the optimum approach for generating practical fusion energy. Stellarators, spherical tori, or a variety of exotic architectures could eventually prove more effective, or economical. However, the tokamak was selected for ITER because it is the most mature and best understood configuration to support an extremely wide field of inquiry for probing, manipulating, and finally understanding the complex dynamics of burning hydrogen plasma. The advances in both theoretical and practical understanding that result from ITER experimental

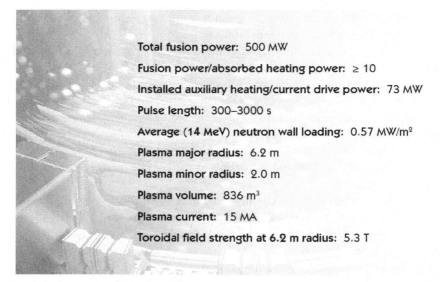

Total fusion power: 500 MW

Fusion power/absorbed heating power: ≥ 10

Installed auxiliary heating/current drive power: 73 MW

Pulse length: 300–3000 s

Average (14 MeV) neutron wall loading: 0.57 MW/m²

Plasma major radius: 6.2 m

Plasma minor radius: 2.0 m

Plasma volume: 836 m³

Plasma current: 15 MA

Toroidal field strength at 6.2 m radius: 5.3 T

Fig. 5.   Major physical characteristics and functional performance parameters of the ITER Tokamak.
Source: US ITER.

research will contribute meaningfully to any future magnetic confinement fusion technology.

7 Major Elements and Distributed Systems

While the complexity and scale of ITER is unprecedented, the subsystems and elements that comprise the integrated system all fall within the realm of either proven technology or incremental advances that can be attained through focused engineering R&D. In this respect, the design and construction of a fully operable ITER tokamak is most appropriately viewed as an engineering challenge, whereas the achievement of quasi-steady-state burning plasma with a net power gain remains the scientific challenge. The following principal subsystems and elements for meeting these challenges will be summarized in the sections that follow.

a. Power supply.
b. Superconducting magnets.
c. Vacuum vessel and internal elements.
d. Cryostat and thermal shield.
e. Fueling pellet injection.

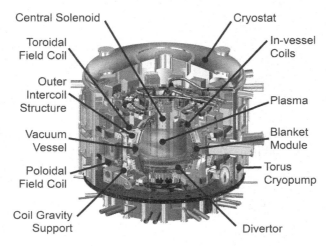

Central Solenoid
Toroidal Field Coil
Outer Intercoil Structure
Vacuum Vessel
Poloidal Field Coil
Coil Gravity Support
Cryostat
In-vessel Coils
Plasma
Blanket Module
Torus Cryopump
Divertor

Fig. 6. CAD drawing of the ITER Tokamak.
Source: US ITER.

f. Plasma heating.
g. Tokamak cooling.
h. Exhaust processing.
i. Biological shield.
j. Controls and instrumentation.

Figure 6 provides a computer-aided-design (CAD) drawing of the integrated ITER tokamak.

8 Power Supply

The ITER facility complex is linked by a 1 km extension to the 400 kV Prionnet Substation operated by Electricity of France (EDF) — the world's largest producer of electricity, which supplies approximately 20% of the electrical needs of the European Union and relies on nuclear fission for over 80% of that supply.[12]

The ITER Steady-State Electrical Network (SSEN) is an AC distribution system, consisting of standard commercial grade AC power system components servicing all conventional building loads with an approximately 120 MW power capacity. The SSEN also includes two diesel generators for emergency backup. Since the SSEN provides site power during the construction phase, it is the first major ITER subsystem to be installed and commissioned.

8.1 *Pulsed load*[13]

Tokamaks operate as pulsed electrical systems with very high cyclical loads. Pulse lengths vary from 300 sec up to an hour, with a nominal 1800 second repetition period. The total load includes power required for:

- Ramping up and sustaining the pulsed magnetic field and plasma current;
- Position and shape control;
- Heating and current drive; and
- Resistive loss compensation.

The loads nominally occur in four phases:

- Pre-magnetization;
- Plasma initiation;
- Plasma steady-state; and
- De-magnetization.

During a nominal plasma pulse, the AC/DC converters generate extremes in reactive power variation where current and voltage are no longer in phase. The short and steep pulses of active and reactive power negatively affect the power quality of the electric network. While the Prionnet Substation can provide up to 500 MW of active power, it has a limited capacity of 200 Mvar for reactive power. As a result, the ITER plant must provide additional reactive power compensation (RPC). The ITER Pulsed Power Electrical Network (PPEN) is designed to address this, as well as provide harmonic filtering (HF). It will have a capacity of 500 MW and be able to compensate 750 Mvar of reactive power. This is accomplished with three RPC and HF units controlling voltage at the 66 kV bus bar level, each having one thyristor controlled reactor (-250 Mvar) and six harmonic filters ($+250$ Mvar in total). Figure 7 depicts the Pulsed AC Distribution Network and Fig. 8 provides a simplified single line diagram of RPC and HF.

The PPEN is also designed to continue normal operations during major transient disturbances. Magnet quench (the transition of the magnet windings from superconducting to normal conducting) represents the worst-case event, since it will lead to large thermal and electromagnetic stresses on the system. In such a case, the load drops out in 50 msec and the PPEN is designed to withstand this transient fifty times over its life cycle.

8.2 *Superconducting magnets*

The ITER electromagnetic field is established by a series of TF coils and further shaped by rings of PF coils with error fields managed by a series of

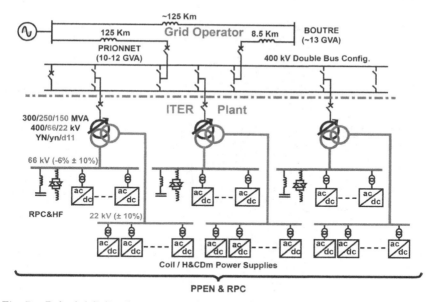

Fig. 7.   Pulsed AC distribution network.
Source: A. D. Mankani, I. Benfatto, J. Tao, J. K. Goff, J. Hourtoule, J. Gascon, D. Cardoso-Rodrigues, and B. Gadeau in IEEE/nuclear and plasma physics society 25th symposium on fusion engineering, 2011, SP#-43.

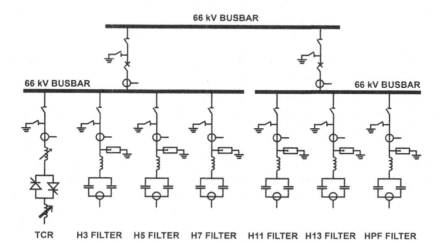

Coil / H&CDm Power Supplies

Fig. 8.   Simplified single line diagram for RPC and harmonic filtering.
Source: A. D. Mankani, I. Benfatto, J. Tao, J. K. Goff, J. Hourtoule, J. Gascon, D. Cardoso-Rodrigues, and B. Gadeau in IEEE/nuclear and plasma physics society 25th symposium on fusion engineering, 2011, SP#-43.

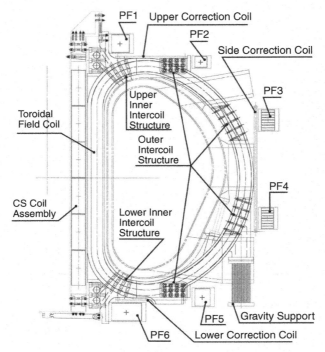

Fig. 9.   Typical cross-section through torus indicating magnet locations.
Source: US ITER.

smaller correction coils (CC). A single large CS coil is employed to induce
and regulate current in the plasma. Figure 9 provides an elevation through
a typical cross-section of the torus indicating the locations of TF, PF, CC,
and CS coils. All magnet structures are designed for 30,000 tokamak pulses
at full field and a 15 mega-amp nominal plasma current.

Superconducting magnets lose electrical resistance when cooled down to
very low temperatures, thus allowing greater electrical efficiency during the high power operations required in tokamaks. This results in an
attractive ratio of power consumption to cost for the long plasma pulses. All
of the large ITER magnets are superconducting, and cooling is achieved by
circulating supercritical helium in the range of 4 Kelvin (−269°C) through
the cores of the cable-in-conduit conductors.

Superconducting cable-in-conduit is formed by first twisting individual
superconducting and copper wire strands together into a bundle. Several
bundles are then interwoven together around a stainless steel tube that

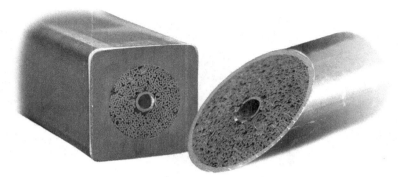

Fig. 10. Typical cross-sections for round and rectangular cable-in-conduit superconductors.

serves as the coolant channel. Finally, this subassembly is inserted inside a stainless steel jacket and compacted. The resulting cable-in-conduit is spooled for shipping to another location where it is de-spooled and run through winding and forming machines that produce magnets in the needed shapes and sizes. Figure 10 indicates typical circular (TF conductor) and rectangular (CS conductor) cross sections for superconducting cable-in-conduit.

9 TF Coils

The primary ITER electromagnetic field for confining the plasma is formed by 18 D-shaped TF coils having field strength of 11.8 Tesla and total magnetic energy of 41 Gigajoules. The TF coil current is 68 kiloamps, with an 11 second discharge time constant, and each coil has a centering force of 403 meganewtons (enough force to lift a 40,000 tonne object). The coils are wound with 115 km of superconducting niobium-tin (Nb_3Sn) conductor and placed in a stainless steel case. Each coil weighs over 350 tons including the structural coil case.

In order to maintain high operational reliability, the TF conductor is embedded in grooved radial plates mounted inside the structural coil case. Since the coil case experiences cyclical loading from the out-of-plane forces generated by interaction of both the TF coil current and PF coil current, a combination of shear keys and pre-compression rings are used to provide a centripetal preload at assembly. Figure 11 illustrates the TF coil location in the tokamak and highlights a single coil-in-case.

The radial plate design was successfully demonstrated during the TF Model Coil Project conducted during the 1990s EDA period. Solutions

Fig. 11.   TF coil positioning in Tokamak highlighting single coil in case.
Source: US ITER.

were also confirmed for issues involving fatigue life of the conductor jacket
and insulation reliability. As a result, the engineering performance and
industrial manufacturing feasibility of the conductor and magnets are well-
established, and the final production TF coils are now in the fabrication
phase for ITER.

10 PF Coils

The six circular and horizontally-positioned PF coils were optimized to provide additional magnetic field control to shape the plasma vertically and
radially, and maintain plasma equilibrium. The coils are wound from superconducting niobium-titanium (NbTi) alloy in square jackets, and normally
operate at 45 kilo-amps with a 14 second discharge time constant. These
coils range from 8–24 m in diameter and are illustrated in Fig. 12 along
with the three sets of CC discussed below.

The six PF coils are attached from top to bottom on the exterior of TF
coil cases by flexible plates that allow for radial displacements. This positioning presents removal and replacement challenges in the event of failure.
All coils include double turn insulation with a metal screen in between that
permits detection of an incipient short prior to full failure. This will allow
disconnection of coil layers and bypassing with bus bar links. The remaining
layers can then be operated in a backup mode at higher current, thereby

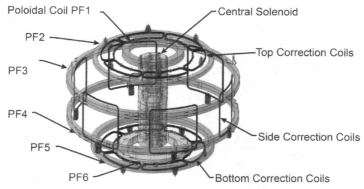

Fig. 12.   PF coil and CC layout.
Source: A. Foussat, P. Libeyre, N. Mitchell, Y. Gribov, C. T. J. Jong, D. Bessette, R. Gallix, P. Bauer, and A. Sahu, *IEEE transactions on applied superconductivity*, **20**, 3 (2010).

reducing the risk of coil repair or replacement. While the top two coils could be removed from the cryostat and repaired or re-wound, the remaining four coils would have to be repaired in place. To further reduce risk, the two central coils, which have the greatest access constraints, include metal plate separators with individual ground insulation between the layers.

10.1 *Correction coils*[14]

The location and geometry of TF, PF, and CS magnet coils will not be precise due to small variations in manufacturing and assembly tolerances. This will lead to variances in the axial symmetry of the magnetic field that can in turn cause locked modes in the plasma and consequent disruptions. Such variances are termed "error fields" and the purpose of the CC magnets is to reduce the range of field imperfections by positioning three sets of six coils each at the top, side, and bottom of the torus. The bottom set has peak field strength of 4 Tesla, while the side and top sets range from 2.26–2.45 Tesla.

Each coil is rectangular and slightly concave similar to an automotive windshield, and constructed from superconducting NbTi cable-in-conduit coil enclosed in a 20-mm thick stainless steel casing. The casing is rigidly connected to the cases of the TF coil set.

10.2 *Central solenoid*

The CS acts as a transformer inducing the majority of the magnetic flux change needed to initiate the plasma, generate the plasma current, and maintain the current during burn time. The CS is made of six independent

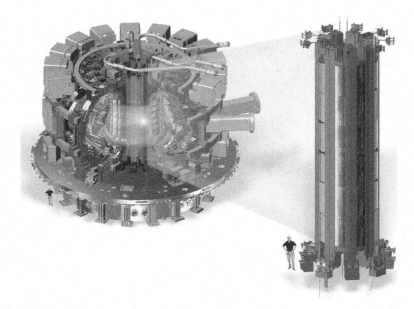

Fig. 13.   Integrated CS with supporting structure.
Source: US ITER.

coil packs that are composed of superconducting Nb_3Sn alloy with each coil pack weighing 110 tons. Approximately 42 km of cable-in-conduit conductor will be used to fabricate the CS. Once integrated it will have peak field strength of 13.1 Tesla, stored energy capacity of 5.5 Gigajoules, and nominally operate at 14 kilovolts and 45 kiloamps. Figure 13 depicts the stack of six CS modules within the support structure.

Each of the six coils consists of 14 turns radially and 40 turns high. Only seven lengths of conductor are used to minimize the use of joints and reduce risk of failure. A superconducting bus bar runs vertically across the outer perimeter to connect each coil pack. The stack is supported from the bottom by the TF coils through a pre-loaded structure consisting of nine internal and 18 external tie-plates. This provides axial pressure on the stack and prevents separation of the modules during operation. Each tie-plate is forged as a single steel member so that when the structure is assembled it can withstand 30 meganewtons of force — equivalent to the force produced by two space shuttles at lift off.

10.3 *Vacuum vessel and internal elements*

The vacuum vessel serves as the plasma chamber and the first containment barrier. Inside the vacuum vessel are internal, replaceable components, including blanket modules, divertor cassettes and port plugs such as the limiter, heating antennae, test blanket modules, and diagnostics modules. These components absorb the radiated heat as well as most of the neutrons from the plasma and protect the vessel and magnet coils from excessive nuclear radiation and heating.

The heat deposited in the internal components and in the vessel is transferred to the environment by means of a cooling water system. It is comprised of individual heat transfer systems. Some elements of these heat transfer systems are also employed to bake and consequently clean the plasma-facing surfaces inside the vessel by releasing trapped impurities. The system is also designed to prevent the possibility of releases of tritium and activated corrosion products to the environment.

The torus-shaped vacuum vessel is located inside the bore of the TF coils and provides the low gas pressure conditions needed to initiate and maintain fusion reaction plasma. In this torus-shaped chamber, the charged plasma particles follow the magnetic field surfaces, thereby avoiding contact with the vessel walls. The magnet system together with the vacuum vessel and internals are supported by gravity supports, one beneath each TF coil.

The ITER vacuum vessel will be twice as large and 16 times as heavy as any previous tokamak chamber, with an internal (minor) diameter of 6 m. It will measure a little over 19 m (major diameter) across by 11 m high, and weigh in excess of 5,000 tons.[15]

The vacuum vessel, illustrated in Fig. 14, will have double-steel walls, with the interspace filled with cooling water. The plasma-facing surfaces of the vessel will support the in-vessel coils and a continuous layer of blanket modules that will capture the escaping fast neutrons generated by the fusion reactions.

Forty-four ports will provide access to the vacuum vessel for remote handling operations, diagnostic systems, heating, and vacuum systems (18 upper ports, 17 equatorial ports, and nine lower ports).

Because the vacuum vessel is the primary containment barrier against release of tritium and activated dust, it is being constructed according to RCC-MR, the French nuclear code. Due to the nuclear material confinement under high water pressure, it must also meet the essential French safety requirements for nuclear pressure equipment (ESPN).[16]

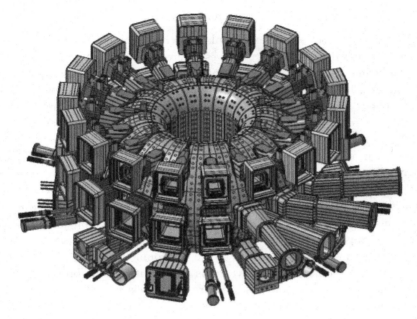

Fig. 14.   Vacuum vessel with port extensions.
Source: ITER organization.

10.4 *Blanket system*

The blanket covers the interior surfaces of the vacuum vessel, providing shielding to the vessel and the superconducting magnets from the heat and neutron fluxes of the fusion reaction. The neutrons are slowed down in the blanket, where their kinetic energy is transformed into heat energy and collected by the cooling water system.

For purposes of maintenance on the interior of the vacuum vessel, the blanket wall is modular. It consists of 440 individual blanket modules, each measuring 1×1.5 m and weighing up to 4.6 tons. Each module has a detachable first wall, which directly faces the plasma to absorb the plasma radiation and charged particle heat load while protecting the vessel from any plasma impingements, plus a semi-permanent blanket shield dedicated to the neutron shielding. The blanket system is illustrated in Fig. 15.

The ITER blanket is one of the most critical and technically challenging components in the ITER system. Together with the divertor, the blanket directly faces the plasma. Because of its unique physical properties, beryllium, a hazardous material requiring special handling, has been chosen

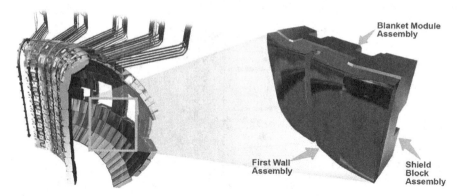

Fig. 15. Blanket module assembly.
Source: US ITER.

to cover the plasma-facing surface "first wall" of the blanket. The rest of the blanket shield layers will be made of high-strength copper backed by stainless steel with internal water-cooling channels.

10.5 *In-vessel coils*[17]

There are two sets of the magnetic coils inside the vacuum vessel — the vertical stability coils and the edge-localized mode (ELM) control coils. The vertical stability coils consist of continuous windings above and below the mid-plane that provide fast vertical position control of the plasma. The ELM coils consist of nine sets of three window-framed coils that produce a resonant magnetic perturbation, which limits the energy in ELM events or suppresses the ELMs altogether. Unmitigated ELMs would cause substantial erosion of the plasma-facing components, especially the divertor. Figure 16 shows the configuration of vertical stabilization and mode suppression coils.

10.6 *Divertor*

Located at the very bottom of the vacuum vessel, the divertor collects and neutralizes the charged particles leaving the plasma — including the inert helium "ash" — and directs these neutral gas particles *via* the vacuum pumping system to the exhaust processing system. The divertor consists of 54 remotely-handled removable cassettes, each holding three plasma-facing component assemblies, or targets, as illustrated in Fig. 17. These are the inner and the outer vertical targets, and the dome. The targets are

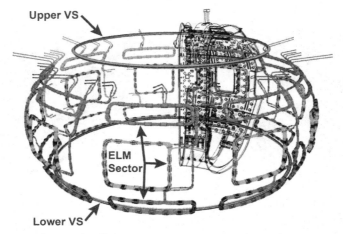

Fig. 16.   Internal vertical stabilization and mode suppression coils.
Source: C. Neumeyer *et al.*, *J. Fusion Science and Technology* (2011), 60.

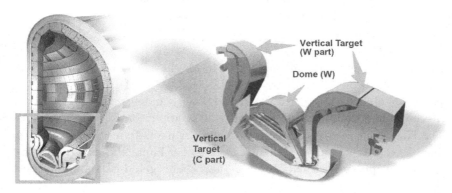

Fig. 17.   Divertor module.
Source: US ITER.

situated at the intersection of magnetic field lines where the high-energy plasma particles strike the components and their kinetic energy is transformed into heat. The heat flux received by these components is extremely intense and requires active water-cooling. The choice of the surface material for the divertor is an important one.[18] Only a few materials are able to withstand temperatures of up to 3,000°C for the projected 20-year lifetime of the ITER machine, including carbon fiber reinforced carbon (CFC) and tungsten; the current choice is tungsten, since it has less propensity to adsorb the tritium fuel.

10.7 *Cryostat and thermal shield*

The cryostat is a large, stainless steel structure surrounding the *tokamak* that provides the insulating vacuum for the superconducting magnets, much like a giant thermos bottle. It consists of a single wall cylindrical construction, reinforced by horizontal and vertical ribs. The cryostat is 29.3 m tall and 28.6 m in diameter.[19]

The cryostat has many openings, some as large as 4 m in diameter, which provide access to the vacuum vessel for cooling systems, magnet feeders, auxiliary heating, diagnostics, and the removal of blanket and divertor components. Large, leak-tight bellows are used between the cryostat and the vacuum vessel to allow for differential thermal contraction and expansion in the structures. Each of these openings has sealed closures to allow total evacuation of the cryostat before commencing operation. The cryostat is completely surrounded by a 2-m-thick concrete biological (bio) shield.

The thermal shield is a set of stainless steel panels, cooled with supercritical helium at 80 K, that provide a thermal radiation barrier between the magnet set and any warmer surfaces (i.e. vacuum vessel and cryostat).[20]

10.8 *Fueling*

The eventual goal for hydrogen fusion is to process tritium in a closed cycle, as illustrated schematically in Fig. 18.

As a first step to starting the fusion reaction, all gases must be evacuated from the vacuum vessel. A vacuum roughing system begins the drawdown, followed by the main pumping system that consists of a set of six torus exhaust cryo-pumps. The cryo-pump panels will be cooled with supercritical helium in order to condense the deuterium and tritium and other gas streams.

Low-density gaseous fuel is then introduced into the vacuum vessel by a gas injection system. Once the fuel is in the vacuum chamber, microwaves are used to pre-ionize the fuel, then an electrical current is applied *via* the CS coil system that completes the electrical breakdown of the gas, initiates the toroidal current, and forms a magnetically confined plasma.

A second fueling system, a pellet injector, will also be used at ITER. The pellet injector operates like a high efficiency icemaker for frozen fuel pellets. An extruder punches out several millimeter-sized deuterium-tritium ice pellets that are propelled by a gas gun at approximately 300 m/s — fast and cold enough to penetrate deep into the plasma core, where they vaporize

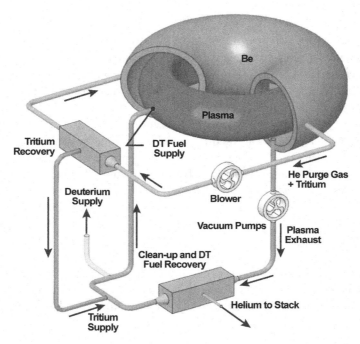

Fig. 18.   Fuel cycle schematic.
Source: ITER organization.

and deposit fresh fuel. The frozen pellets are injected through a set of guide tubes located in the inner wall of the vacuum vessel and another guide set of tubes for outer wall injection. Prototype pellet injectors, illustrated in Fig. 19, are being developed at Oak Ridge National Laboratory.

Pellet injection is the principal tool used to control plasma density and is also efficient at controlling Edge Localized Modes a (ELMs). Injecting small frozen deuterium pellets in the edge plasma, has been shown to be effective in ELM mitigation.[21]

Less than 1 g of fusion fuel is present in the vacuum vessel at any moment in time. The divertor, located at the bottom of the vacuum vessel, permits recycling of any fuel that is not consumed. Unburned fuel flows to the divertor, is pumped out and separated from helium produced during the fusion reaction, mixed with fresh tritium and deuterium, and is re-injected into the vacuum chamber.

Fig. 19.   Prototype pellet injector.
Source: US ITER/ORNL.

10.9 *Plasma heating*

Due to the absence of any equivalent to the sun's gravity, to achieve sufficient pressure, the plasma temperature inside the ITER tokamak must reach greater than 150 million°C, or 10 times the temperature at the core of the Sun, in order for the gas in the vacuum chamber to reach the plasma state and for efficient fusion reactions to occur. The hot plasma must then be sustained at these extreme temperatures in a controlled way in order to extract net energy.

ITER will rely on three sources of external heating which will work in concert to provide the required input plasma heating power of 50 MW: neutral beam injection and two sources of high-frequency electromagnetic waves.[22] The characteristics of these systems are listed in Fig. 20.

Heating System Characteristic	Neutral Beam (NB)	Electron Cyclotron (EC)	Ion Cyclotron (IC)
Energy or frequency	1 MeV	17 GHz	40–55 MHz
Power injected-per unit equatorial port (MW)	16.5	20	20
Number of units for the first phase	2	1	1
Total Power (MW) for the first phase	33	20	20

Fig. 20. Heating and current drive systems.

Source: US ITER.

10.10 *System cooling*

ITER will be equipped with a cooling water system to manage the heat generated during operation of the *tokamak*. The internal surfaces of the vacuum vessel (first wall blanket and divertor) must be cooled to less than 600°C, only a few meters from the 150-million-degree plasma.

Pressurized water will be used to remove heat from the vacuum vessel and its components, and to cool auxiliary systems such as radio frequency heating and current drive systems, the chilled water system, the cryogenic system, and the coil power supply and distribution system. The cooling water system incorporates multiple closed heat transfer loops plus an open-loop heat rejection system (HRS). Heat generated by escaping fast neutrons slowing down in the vacuum vessel components during the deuterium–tritium reaction will be transferred through the primary cooling water system to the intermediate component cooling water system, and to the HRS, which will reject the heat to the environment. The cooling water system must reject over 1 GW of thermal energy.

10.11 *Biological shield*[23]

Close attention to radiation dosage rates in the tokamak building is critical to ensure occupational safety. The cryostat is surrounded by a two-meter thick concrete bio-shield as the primary means of attenuating radiation loads in plant areas that require human activity. The cryostat includes

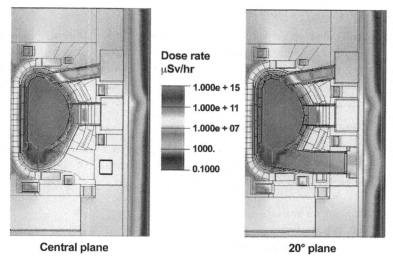

Central plane   **20° plane**

Fig. 21.   Dose rates at two planes in ITER.
Source: A. M. Ibrahim, M. E. Sawan, S. W. Mosher, T. M. Evans, D. E. Peplow, P. P.
Wilson, and J. C. Wagner, *Fusion Science Tech.* **60**, 2 (2011), p. 676–680.

numerous ports for diagnostics, heating, plasma exhaust, etc. that represent
strong sources of streaming neutrons and photons. Full three-dimensional
nucleonics modeling is therefore necessary to address the geometric details
associated with such a large and complex structure.

Figure 21 shows a dose rate map for the central plane and on a plane
rotated at 20° from the central plane. The central plane bisects the center
of an upper diagnostics port and an equatorial port, while the rotated plane
bisects two similar ports as well as a divertor port for plasma exhaust. This
approach effectively characterizes the areas of peak flux.

The dose rate varies due to use of shielding plugs of varying thickness
at the diagnostic ports, as well as gaps between the port walls and shielding
plugs. The divertor pumping ports do not employ shielding plugs and therefore strongly affect the dose rate at the bio-shield, as indicated in the lower
portion of 20° plane map. Analysis has shown that the bio-shield reduces
the total prompt operational dose by six orders of magnitude. The peak
values of the prompt dose rates at the back surface of the bio-shield were
240 μSv/hr and 94 μSv/hr corresponding to the regions behind the divertor port and the equatorial port, respectively. For comparison purposes,
the United States Code of Federal Regulations (10CFR20 and 10CFR835)
limits radiation worker total annual exposure to 50 mSv, or 50,000 μSv.

10.12 *Instrumentation & controls*

The Instrumentation & Control (I&C) system for ITER relies on a vast array of diagnostic sensors, analytical control schema software and actuators designed to operate the tokamak safely and enable the next generation of R&D for burning plasma.

10.13 *Diagnostic instruments*[24]

The ITER I&C system will employ a wide number of individual measuring systems that have been drawn from the full range of modern plasma diagnostic techniques, including:

- Plasma position reflectometry.
- Residual gas analysis.
- Neutron flux monitoring and particle analysis.
- Continuous external Rogowskis.
- Neutral particle analysis.
- Gamma ray spectrometry.
- Visible and infrared imaging.
- Discrete inductive sensing.
- Microfission chambers.
- Neutron activation systems.
- Edge imaging X-ray crystal spectroscopy.
- High and low field side reflectometry.
- Pressure gauges.
- X-ray crystal spectroscopic surveying.
- Fiber optic current sensing.
- Thermocouples and thermography.
- H-Alpha and visible spectrometry.
- Bolometry.
- Charge exchange recombination spectrometry.
- Motional Stark effect polarimeter.
- Collective and edge Thomson scattering.
- Langmuir probes.
- Beam emission spectrometry.
- Polarimetry.
- Impurity monitoring.
- Neutron and X-ray cameras.
- Electron cyclotron emission.

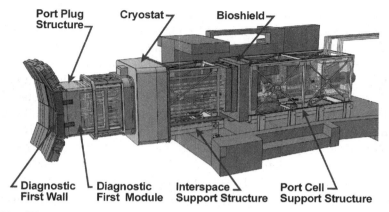

Fig. 22.   Diagnostic instrument location in ports.
Source: D. Johnson *et al.*, in 20th Topical Conference on High-Temperature Plasma Diagnostics, 2014.

- Interferometry.
- Vacuum ultraviolet surveying, edge imaging, and spectrometry.

Many instruments will be located in ports and constructed as standardized modules that extend from the plasma "first wall" through the vacuum vessel, cryostat, interspace, and bio-shield as illustrated in Fig. 22.

Because of the harsh environment inside the vacuum vessel, these systems will have to cope with a range of phenomena not previously encountered in diagnostic implementation, all while performing with great accuracy and precision. The levels of neutral particle flux, neutron flux and fluence will be respectively about 5, 10, and 10,000 times higher than the harshest conditions experienced in today's magnetic fusion machines, while the pulse duration will be about 100 times longer. To ensure reliable control and measurements, each desired quantity is measured using two or more independent diagnostic techniques.

10.14 *Control, Data Access, and Communication (CODAC)*[25]

Unusually high energy level, heat flow, neutron flux and long pulse duration combine to create challenging conditions for control of burning plasma. The plasma control system (PCS) will communicate with at least 45 unique diagnostic packages and 20 different actuator systems to sense and respond to rapidly evolving conditions occurring in the plasma stream. The various magnetic coils, gyrotron heaters, coolant pumps, and associated subsystems are also active elements in the overall control schema.

Plasma dynamics can be viewed as a composite of complex large-scale flows, turbulent small-scale flows, and energetic particle interactions. Due to the complexity and uncertainty of responses from the plasma to active control actions, the PCS must be highly robust and reliable in terms of protecting the plant investment during experimental operations. Understanding of reaction dynamics will systematically progress over time through the planned research agenda. The frequency and number of pulse cycles therefore becomes a key factor in the rate of understanding and evolution of steady-state operating principles. Approximately 10,000–12,000 pulses have been projected for the early hydrogen/helium phase of plant commissioning; performance characterization is based on an assumption of dual-shift research operations.

Two separate types of control logic will ultimately be required for ITER: (a) continuous control and (b) exception forecasting, detecting, and handling. Continuous control requires the development of algorithms that can reliably produce feedback to regulate and maintain a nominal operating scenario, while exception handling control demands effective responses to off-nominal and fault conditions, both of which are likely during experimental operations.

The fusion research community has had a strong and significant R&D program underway involving smaller scale tokamak reactors deployed around the world. This work provides a sound foundation upon which to build the scientific understanding and technological control that will result from the scaled up ITER R&D opportunity.

11 Research Plan[26]

The following lists of objectives and goals by phase of operations are extracted from the ITER Research Plan. Research operations are divided into four phases associated with a gradual progression from initial characterization and validation of the performance attributes of the tokamak system to full-up operation in a burning plasma state that allows achievement of mission objectives. The four phases include:

1. Hydrogen and Helium (HH) Phase.
2. Deuterium (D) Phase.
3. Deuterium-Tritium Phase 1 (DT1).
4. Deuterium-Tritium Phase 2 (DT2).

Hydrogen and Helium Phase: The overall objectives for the hydrogen/helium phase of ITER operation are to:

- Establish routine operation of the tokamak and its subsystems with plasmas;
- Commission all installed heating and diagnostic systems with plasma;
- Commission installed fueling systems with plasma;
- Commission and integrate all installed control systems (including in-vessel coil systems);
- Commission and integrate all safety related systems;
- Demonstrate plasma operation to full technical performance;
- Perform validation of diagnostic data and demonstrate consistency of measurements;
- Characterize aspects of plasma performance critical for subsequent phases of operation;
- Characterize operational boundaries and off-normal events;
- Demonstrate reliable avoidance or mitigation of off-normal events;
- Validate licensing assumptions concerning disruptions;
- Characterize hydrogenic retention and demonstrate techniques to be used later for control of tritium inventory;
- Establish and characterize type-I ELMy H-modes, most likely in helium plasmas, and demonstrate ELM mitigation/suppression;
- Demonstrate, to the extent possible, plasma performance and scenarios envisaged for D and DT, including plasma operation on tungsten plasma facing components;
- Conduct first exploration of fusion plasma physics at ITER scale and parameters.

Deuterium Phase: The primary goals of the deuterium phase are to:

- Commission the heating and current drive (CD) systems to the level of performance and reliability required for a successful DT program;
- Establish reliable, long-pulse plasma operation on tungsten divertor targets in ohmic, L- and H-mode;
- Develop, demonstrate, and validate H-mode scenarios up to the highest parameters achievable and, if the H-mode scaling turns out to be favorable, to full machine parameters;
- Demonstrate that hydrogenic retention in H-mode is acceptable for DT operation;

- Demonstrate ELM amelioration sufficient for divertor protection over the expected operational range;
- Commission and validate H-mode relevant diagnostics (e.g. neutron detectors, pedestal diagnostics).

D Phase 1: During the DT1 phase ITER should achieve extended burn in inductively driven plasmas with $Q \geq 10$ for a range of operating conditions, and a duration sufficient to achieve stationary conditions on the time scales characteristic of plasma processes. More detailed ITER operational objectives during the DT1 inductive and non-inductive phases are as follows.

Inductive plasmas:

- Develop burn control techniques for DT plasmas, including power and particle exhaust, active MHD control;
- Achieve fusion power of several hundred MW;
- Demonstrate $Q \geq 10$ for several hundred seconds;
- Develop a hybrid mode of operation for longer burn durations or higher fusion performance to the extent possible;
- Pursue a program of burning plasma research based on $Q \geq 10$ operating scenarios.

Non-inductive plasmas: The objective of the DT1 Phase will be to develop — to the maximum extent possible — the basis for non-inductive plasma operation toward the ultimate moderate-Q, steady-state operation. It is likely that this goal will not be fully accomplished until the DT2 Phase:

- Extend current drive studies to DT plasmas — quantify off-axis CD capability;
- Build on the DT-inductive program by establishing a range of target q-profiles with early heating and current ramps;
- Commission feedback control algorithms for H&CD, MHD stability control, fueling, and divertor power handling in relevant regimes;
- Explore control algorithms in the presence of strong heating over the current relaxation time, and validate the models for these control algorithms;
- Develop scenarios close to MHD limits and explore stability/control;
- Develop fully non-inductive plasmas and extend performance to $Q = 5$;
- Extend pulse length towards 3000s with $Q \geq 5$;
- Pursue burning plasma physics studies in non-inductive scenarios.

Towards reactor-relevant plasmas and technology: during and subsequent to the development of the operation that achieves these objectives, it is envisioned that experiments will be conducted to explore a wide range of plasma physics issues in the burning plasma state. However, the extended research plan might require additional time for the physics research program in DT2. Key scientific objectives include:

- Improve the understanding of plasma physics phenomena in reactor conditions;
- Validate theory- and simulation-based predictive models of key performance related phenomena;
- Demonstrate the capability for using model-based tools for controlling primary aspects of reactor-grade plasma.

DT Phase 2

The present version of the Research Plan is restricted to identifying possible research priorities for DT2. These include:

- The test blanket module development program;
- Full steady-state demonstration with additional heating and current drive tools;
- Extension of ITER regimes towards those required for a reactor, e.g. higher β and higher radiated power fraction;
- Demonstration of compatibility of operation regimes with a DEMO-relevant wall;
- Demonstration of potential DEMO regimes with a reduced number of heating and current drive systems;
- Demonstration of the plasma control required for DEMO using only DEMO-compatible diagnostics.

12 Safety and Licensing

Safety is a top-priority issue for the project and includes consideration of the safety of the project staff and workers on site, the local population and the environment. French nuclear regulations have been applied throughout the design phase of the project, and will continue to be followed during construction, operation, and decommissioning.[27]

The fusion process itself is inherently safe. In a tokamak fusion device, the quantity of fuel present in the vessel at any one time is sufficient for a few-seconds burn only. It is difficult to reach and maintain the precise

conditions necessary for fusion; any significant degradation of these conditions will cause the plasma to cool within seconds and stop the reaction.
There is no danger of a run-away reaction, because fusion does not involve
a self-perpetuating chain reaction unlike fission.

When the highly energetic neutrons interact with the walls of the internal components and the plasma chamber, these materials become activated.
In-vessel materials can also become contaminated with small amounts of
tritium and radioactive dust composed mainly of beryllium and tungsten.

In ITER, confinement of these materials will be based on the principle of defense-in-depth — materials with the highest radioactive content
are located in the very center, surrounded by multiple protective layers.
Maintenance and refurbishment of the more radioactive elements and components of the tokamak are performed using machines and tools controlled
remotely to avoid human exposure to radioactivity. As previously discussed,
2-m-thick protective concrete walls serving as a bio-shield completely surround the tokamak.

During the operational lifetime of ITER, remote handling will be used
to refurbish components of the vacuum vessel. All waste materials will be
treated, packaged, and stored on site in a Hot Cell building to maintain total
separation. The half-life of most radioisotopes contained in this waste is less
than 10 years. The fusion reaction will produce no long-lived waste. Within
100 years, the radioactivity of the materials will have diminished to such a
degree that the materials can be recycled for use in future fusion plants.[28]

The confinement of tritium within a closed fuel cycle is one of the most
important safety objectives at ITER, because although tritium has a relatively short radioactive half-life of 12.3 years it nonetheless possesses a
high radio-toxicity. The total amount of tritium present on site will have
a licensed limit of 4 kg. A multiple-layer barrier system has been designed
to protect against spread or release of tritium. The first level of the safety
confinement barrier is the vacuum vessel itself. Inside this double-steel container, the fusion reaction takes place within a near-vacuum. All pumps,
pipes, valves, and instruments leading into the vacuum vessel are highly
leak-tight.

Surrounding the first confinement system is a second level of security
comprising all vessels or systems that surround the vacuum vessel, including
buildings as well as advanced detritiation systems for the recovery of tritium
from gas and liquids. In ITER, these highly developed detritiation systems
will work efficiently to keep the fusion fuels recycled within a closed system and maintain any releases well below regulatory limits. These systems

have been designed to remove tritium from liquids and gases for reinjection into the fuel cycle. Remaining effluents will be well below authorized limits. Gaseous and liquid tritium releases to the environment from ITER are predicted to have a dosage rate below 10 μSv per year. This is 1,000 times lower than ITER's General Safety Objective of 10 mSv per year (the regulatory limit in France). Scientists estimate exposure to natural background radiation to be approximately 6,200 μSv per person per year.

To mitigate seismic risks, the ITER tokamak complex is constructed on a foundation of specially reinforced concrete, and will rest upon bearing pads on top of pillars that are designed to reduce the impact of earthquakes. Cadarache, France is classified as an area of moderate seismic activity. The facility will be equipped with seismic sensors around the site to record all seismic activity, however minor.

ITER safety processes are in full compliance with French and international regulations, and the ITER installation is classed as a "basic nuclear installation" by French authorities. A successful Public Enquiry was held in 2011, and on 20 June 2012 the ITER Organization was informed in writing by the French Nuclear Safety Authority (Autorité de Sûreté Nucléaire) that — following an in-depth technical inspection — the operational conditions and the design of ITER as described in the ITER safety files fulfilled expected safety requirements. As part of its responsibilities as a nuclear operator, the ITER Organization will perform regular checks on the installation during construction and operation. French nuclear authorities will also audit and inspect the ITER Organization's application of regulations.

13 The Coming Era of Burning Hydrogen Plasma

In 2015, the ITER tokamak complex began rising up out of the ground in southern France. Completion of the initial R&D facility is projected to require approximately 10 years. Additional capabilities for DT research operations will follow. The timeline is affected by the rate of annual funding among the seven ITER Partners. Accelerations or delays, both of which are possible, depend on socioeconomic conditions around the world.

The achievement of a capability to sustain burning plasma under short-term, quasi-steady-state conditions is the effective "turning point". At this critical historic juncture, the proven, very large, net energy gain will form a compelling basis for follow-on investments in next-generation containment materials and lithium-to-tritium transmutation techniques necessary to enable long-term, steady-state operations. This technology pull is already underway, but the pressure can be expected to increase in order to seize the commercial opportunity demonstrated by burning hydrogen plasma.

In the global fusion physics community, the step following ITER is envisioned to be state-sponsored, commercial-scale, electric-generating prototypes termed "DEMOs," and the expectation is that the current ITER partners will each independently pursue national DEMOs. At this stage, levels of government, industry and academic participation will likely vary according to national socioeconomic policies. China, Japan, and Korea have initiated planning for DEMOs, and are already investing in upgraded tokamak laboratories and infrastructure (e.g. supercomputers and superconducting cable production).

The power of the stars could achieve practical realization by the mid-21st century. There is little remaining doubt in the informed fusion physics and engineering communities that controlled nuclear fusion for electric generation can be achieved; the only uncertainties are when, at what cost, and by whom? If the paradigms of history are in any way prophetic, then the nations that lead in hydrogen fusion "know how" will likely become the leaders of our world in the future. Energy, a fundamental factor of production, will be available to all due to the ubiquity of fusion fuel; however, the science and technology prowess to "create stars" must be sponsored by national leaders with vision.

Acknowledgments

This manuscript has been compiled and written by UT-Battelle, LLC under Contract No. DE-AC05-00OR22725 with the U.S. Department of Energy. This compilation was based largely on information made available in the public domain by the ITER Organization and US ITER Project Office. In many cases, the technical literature was consulted for details and these references are specifically identified where applicable. The ITER design remains subject to further refinements as the Project continues to evolve. While design details were accurate at the time of writing, some changes are to be expected. This document was prepared by staff in the US ITER Project Office at Oak Ridge National Laboratory and was supported by the U.S. Department of Energy, Office of Science, Fusion Energy Sciences. The document will be maintained by the US ITER Project Office in the public domain at https://www.usiter.org. The site will be updated as necessary to preserve technical accuracy as the ITER Project progresses.

The United States Government retains, and the publisher by accepting the article for publication, acknowledges that the United States

Government retains a non-exclusive, paid-up, irrevocable, world-wide license to publish or reproduce the published form of this manuscript, or allow others to do so, for United States Government purposes. The Department of Energy will provide public access to these results of federally sponsored research in accordance with the DOE Public Access Plan (http://energy.gov/downloads/doe- public-access-plan). The views and opinions expressed herein do not necessarily reflect those of the U.S. Department of Energy.

References

1. U.S. National Academy of Engineering, *Grand Challenges for Engineering* (National Academies of Science Press, 2008).
2. D. A. Petti, B. J. Merrill, J. P. Sharpe, L. C. Cadwallader, L. El-Guebaly, and S. Reyes (2007). Recent accomplishments and future directions in the US Fusion Safety and Environmental Program. *Nucl. Fusion*, **47**(7), S427–S435.
3. P. Z. Grossman (2013). *U.S. Energy Policy and the Pursuit of Failure*. Cambridge University Press.
4. W. M. Stacey (2010). *The Quest for a Fusion Energy Reactor: An Insider's Account of the INTOR Workshop*. Oxford University Press.
5. Executive Summary of the IAEA Workshop, mid-1985 to 1987 (1988). International tokamak reactor group. *Nucl Fusion*, **28**(4), (1988), pp. 711–743.
6. U.S. National Research Council Burning Plasma Assessment Committee (2004). *Burning Plasma: Bringing a Star to Earth*. National Academies Press.
7. IAEA Information Circular. *Agreement on the Establishment of the ITER International Fusion Energy Organization for the Joint Implementation of the ITER Project* (INFCIRC/702, 2007).
8. https://en.wikipedia.org/?title=Tokamak, accessed June 19, 2015.
9. http://www.tokamak.info/, accessed June 22, 2015.
10. Controlled fusion power was first demonstrated at the 10.7 MW level at the Princeton University Tokamak Fusion Test Reactor in 1994 and replicated in 1997 at the 16 MW level (current world record) at the Joint European Torus operated by Culham Science Center owned and operated by the United Kingdom Atomic Energy Authority outside Oxford, United Kingdom.
11. *ibid* (6), p. 2.
12. https://en.wikipedia.org/wiki/Électricité_de_France, accessed June 22, 2015.
13. A. D. Mankani, I. Benfatto, J. Tao, J. K. Goff, J. Hourtoule, J. Gascon, D. Cardoso-Rodrigues, and B. Gadeau, The ITER Reactive Power Compensation and Harmonic Filtering (RPC & HF) System: Stability & Performance. in IEEE/Nuclear and Plasma Physics Society 25th Symposium on Fusion Engineering, 2011, SP#-43.
14. A. Foussat, P. Libeyre, N. Mitchell, Y. Gribov, C. T. J. Jong, D. Bessette, R. Gallix, P. Bauer, and A. Sahu (2010). Overview of the ITER correction coils design. *IEEE Transactions on Applied Superconductivity*, **20**(3).

15. C. Sborchia *et al.* (2013). Design and manufacture of the ITER Vacuum Vessel. in *25th Symposium on Fusion Engineering*, pp. 1–8.
16. H. J. Ahn *et al.* (2011). Fabrication design and code requirements for the ITER vacuum vessel. in *American Society of Mechanical Engineers Pressure Vessels and Piping Conference, Vol 1: Codes and Standards*, pp. 275–283.
17. C. Neumeyer *et al.* (2011). Design of the ITER in-vessel coils. *J. Fusion Science and Technology*, 60.
18. T. Hirai *et al.* (2012). ITER tungsten divertor design development and qualification program. in *Proceedings of the 27th International Symposium on Fusion Nuclear Technology*, pp. 1798–1801.
19. B. Doshi *et al.* (2013). Design and manufacture of the ITER cryostat. in *25th IEEE Symposium on Fusion Engineering*, pp. 1–6.
20. C.H. Noh *et al.* (2012). Final design of ITER vacuum vessel thermal shield. in *Proceedings of the 27th International Symposium on Fusion Nuclear Technology*.
21. L. R. Baylor *et al.* (2009). Pellet fuelling, ELM pacing and disruption mitigation technology development for ITER. *Nucl. Fusion*, **49**, 085013.
22. F. Wagner *et al.* (2010). On the heating mix of ITER. *Plasma Phys. Control. Fusion*, **52**, 124044.
23. A. M. Ibrahim, M. E. Sawan, S. W. Mosher, T. M. Evans, D. E. Peplow, P. P. Wilson, and J. C. Wagner (2011). Global evaluation of prompt dose rates in ITER using hybrid monte carlo/deterministic techniques. *Fusion Science Tech.*, **60**(2), 676–680.
24. D. Johnson *et al.* (2014). ITER diagnostic development. in 20th *Topical Conference on High-Temperature Plasma Diagnostics*.
25. D. Humphreys (2015). Novel aspects of plasma control in ITER. *Phys Plasmas*, **22**, 021806.
26. The ITER Research Plan. ITER Organization, IDM UID-2FB8AC, Nov. 2009.
27. C. Alejaldre, J. Elbez-Uzan and L. Rodriguez (2012). Feedback of the licensing process of the first nuclear installation in fusion, ITER. in *24th IAEA Fusion Energy Conference*.
28. S. Rosanvallon, B. C. Na, M. Benchikhoune, J. Uzan-Elbez, O. Gastaldi, N. Taylor, and L. Rodriguez (2010). ITER waste management. *Fusion Engineering and Design*, **85**, 10–12, 1788–1791.

About the Contributors

Bertrand Barré (Chapters 1–3) is Professor Emeritus of Nuclear Engineering at the French *Institut National des Sciences et Techniques Nucléaires* (INSTN), and a teacher at Sciences Po/Paris School for International Affairs. Born in December 1942, B. Barré joined the French Atomic Energy commission, CEA, in 1967 and has been working ever since, both in France and abroad, on the development of Nuclear Power. Alternating scientific and managerial positions, Prof. Barré was notably Nuclear Attaché at the French Embassy in Washington (USA), Director of Engineering in TECHNICATOME (now AREVA-TA), Director of the Nuclear Reactor Directorate of the CEA and Vice-president in charge of R&D in COGEMA (now AREVA-NC). Until December 2012, he was Scientific Advisor to the Chairman of the AREVA group. Bertrand Barré is a Fellow of the American Nuclear Society (ANS), former Chairman of the French Nuclear Energy Society (SFEN), former President of the European Nuclear Society (ENS) and the International Nuclear Societies Council (INSC), former Chairman of the International Nuclear Energy Academy (INEA), and former Vice-Chairman of the Scientific and Technical Committee of EURATOM. He was appointed the first Chairman of SAGNE (Standing Advisory Group for Nuclear Energy) by the Director General of the International Atomic Energy Agency IAEA.

Robert J. Goldston (Chapter 6) is a Professor in the Department of Astrophysical Sciences at Princeton University. He was Director of the DOE Princeton Plasma Physics Laboratory from 1997–2009. He has made numerous experimental and theoretical contributions to fusion plasma physics over the last 40 years, exploring plasma heating, energy confinement, and — since stepping down as Director of PPPL — the physics of plasma power and particle exhaust. He is associated faculty with the Princeton Environmental Institute, and affiliated faculty with Princeton's Program in Science and Global Security, and has been examining socio-economic aspects of fusion power. He is a fellow of the American Physical Society and recipient of its "Excellence in Plasma Physics" award. He also chaired the American Physical Society's Physics Policy Committee. Prof. Goldston is the author or coauthor of many peer-reviewed publications, and co-author of the textbook *Introduction to Plasma Physics* (1995).

John Lindl (Chapter 4) received his B.Sc. in Engineering Physics from Cornell University, USA, in 1968, and his Ph.D. in Astrophysical Sciences from Princeton in 1972. Dr. Lindl is a Fellow of the American Physical Society and of the American Association for the Advancement of Science. In 1993, he was awarded the Edward Teller Medal, and in 1994, he received the E.O. Lawrence Award for work in Inertial Fusion. He is the recipient of the 2000 Fusion Power Associates Leadership Award and received the Maxwell Prize from the American Physical Society in 2007. Dr. Lindl joined Lawrence Livermore National Laboratory, USA, in 1972 as a physicist in A-Division's X-group, concentrating on fluid instabilities and high gain inertial confinement fusion (ICF) targets. In 1978, he became leader of the Laser Target Design Group and in 1981, an associate division leader for the Laser Target Design Group in X-Division. In 1983, Dr. Lindl was named X-Division Leader and Associate Program Leader for Theory and Target Design in the ICF Program. In 1989, he became Deputy Program Leader for Theory and Design and ICF Applications. In 1990, he became ICF Target Physics Program Leader and in 1994, he became the ICF Scientific Director.

His work in ICF has spanned a wide range of topics including high gain target designs for lasers and particle beams, hydrodynamic instabilities in ICF, implosion symmetry and hohlraum design, high energy electron production and plasma evolution in hohlraums, and the physics of compression and ignition. In 1990, Dr. Lindl was placed in charge of the Nova Laser Program with the goal of developing the physics basis for proceeding with the NIF. This work has been the subject of two major review articles written by Dr. Lindl, the first in 1995 following the declassification of much of radiation driven ICF in 1993, and the second in 2004, summarizing the physics basis for proceeding to ignition experiments with indirect drive. A book written by Dr. Lindl, titled *Inertial Confinement Fusion: The Quest for Ignition and Energy Gain Using Indirect Drive* (1998), has become a widely used reference for the science of inertial fusion. In 1999, LLNL integrated its Magnetic and Inertial Fusion Energy Programs and in 2000, Dr. Lindl became the first leader of this combined research program, a position he held until 2004. In October 2004, he was selected as a Teller Fellow by the LLNL Director to concentrate on the NIF ignition program and in 2010, Lindl was selected as one of the first three Distinguished Members of the Technical Staff (DMTS) at LLNL. From 2006 to 2014, Dr. Lindl served as the NIF Programs Chief Scientist with the primary goal of achieving ignition on the NIF. In 2014, Dr. Lindl was the lead author on a review paper that provided a comprehensive summary of results achieved during the course of the National Ignition Campaign (NIC) as well as the path forward currently underway to address the remaining challenges on the path to ignition.

John Sethian (Chapter 5) recently retired from the Naval Research Laboratory, USA, as head of the Electron beam Science and Applications Section of the Laser Plasma Branch in the Plasma Physics Division. He has worked as a scientist at the Naval Research Laboratory (NRL) since 1977. At NRL, Dr. Sethian worked on a broad range of topics in plasma physics, electron beam physics, pulsed power, and lasers. The underlying theme of his research has been the development the science and technologies needed for a practical fusion power source. He has worked on all approaches to fusion energy: inertial, magnetic, and z-pinch based. His most recent contribution was founding and directing the national "High Average Power Laser" (HAPL) Program. This program was directed to develop the technological underpinnings for practical fusion power based on lasers and the direct drive. The program brought together more than 60 researchers from national labs, universities, and private industries. The key science and technologies were developed in concert with one another as a part of a coherent system. Credible solutions for almost all the key components were developed. As part of the HAPL program, Dr. Sethian developed the science and technologies needed to build a repetitively pulsed, large area, high energy electron beam source. He first applied the technology to develop a durable and efficient electron beam pumped krypton fluoride (KrF) Laser to meet the requirements for fusion energy. He later applied the technology to a wide range of applications, including material surface modification, fuel reformation, and most recently, the elimination of NO_x in fossil fuel power plants. Dr. Sethian was born in Washington, DC, USA, and attended public schools in Arlington County, VA. He received an A.B. degree in Physics from Princeton University in 1972, and a Ph.D. in Applied Physics from Cornell University, USA, in 1976. He is a Fellow of the American Physical Society, has received four NRL invention/technology transfer awards, four patents, three NRL publication awards, and has published over 80 archival papers. He has received the Fusion Power Associates Leadership Award, the American Nuclear Society's Annual Outstanding Achievement Award, and the US Navy Meritorious Service Medal.

Erik Storm (Chapter 4) is internationally recognized as an expert in the physics and technology of Inertial Confinement Fusion (ICF). He received a B.Sc. in 1967, M.Sc. in 1968, and Ph.D. in 1972, in Aeronautical Engineering and Physics of Fluids, from the California Institute of Technology. He began his career at the Lawrence Livermore National Laboratory (LLNL), USA, in 1974, and during the next 20 years worked on, and directed, the ICF campaigns on the Janus, Argus, Shiva and Nova facilities at LLNL, becoming the Deputy Associate Director for Lasers in 1985. Over the next 8 years he was responsible for executing a series of dedicated underground nuclear tests that laid to rest the question of the scientific feasibility of ICF, and led the LLNL ICF Program through two National Academy of Sciences (NAS) reviews that culminated in the NAS recommendation to build the National Ignition Facility (NIF). In 1994 he was asked to negotiate and put in place a Government-to-Government agreement between the US Department of Energy and the French *Commissariat de l'Energie Atomique* to collaborate on High Energy/High Power lasers. This agreement allowed an extremely fruitful collaboration between the two countries to develop the technology for, and subsequently build, the NIF at LLNL, and the Laser Megajoule (LMJ) at the CEA laboratory near Bordeaux, France. Dr. Storm returned to LLNL as Deputy Associate Director for the Defense Nuclear Technologies program from 1997–1998, before once again taking on the responsibility of coordinating the joint US–French NIF–LMJ technology and physics development programs. From 2006, Dr. Storm has been responsible for developing Advanced Inertial Fusion Energy options at LLNL.

Mark Uhran (Chapter 7) spent 28 years working on design, operations and utilization of the International Space Station — a global partnership between Canada, Europe, Japan, Russia and the United States. During that period, his responsibilities grew to senior executive oversight of cost, schedule and technical scope for the $60 billion, 500 metric ton laboratory complex that has been operating safely and productively in low-Earth orbit with a permanent crew of six since 2009. He retired as Director of the International Space Station Division at NASA headquarters in July 2012, and joined Oak Ridge National Laboratory in August 2012 to manage strategic communications for the US ITER Project — another global partnership. ITER is designed to demonstrate the capability to sustain burning plasma formed by the fusion of hydrogen isotopes, and prove the potential for electric power generation through nuclear fusion. The 500-megawatt thermal industrial-scale experimental reactor is now rising up out of the ground in southern France. Mr. Uhran has over 30 years practical experience in developing complex components and systems that enable experimental laboratories to conduct leading cutting-edge research & development on high risk/high payoff technologies. He is a vocal advocate for applying the fundamental principles of systems engineering and risk management to actively control cost, schedule and technical scope in the pursuit of very large-scale international systems integration projects. Mark Uhran holds a B.Sc. in Natural Resources from Cornell University (1976), USA, an M.Sc. in Technology Management from the University of Maryland (1988), USA, and an M.A. in Public Administration from Harvard University (1998).

Michael Zarnstorff (Chapter 6) is the Deputy Director for Research at the Princeton Plasma Physics Laboratory. He is an experimental plasma physicist with interests in the basic physics of plasma confinement and configuration optimization. Dr. Zarnstorff received his Ph.D. in Physics (1984) from the University of Wisconsin–Madison, USA, and was named a Distinguished Research Fellow by the laboratory in 1995. In 2008, he received the APS Dawson Award for Excellence in Plasma Physics Research. His research included the first observation and systematic study of the bootstrap current, investigations of neoclassical and turbulent transport, transport barriers, and the confinement and stability of different magnetic field configurations. He led the National Compact Stellarator Experiment physics group and was one of the leaders of the TFTR research program. He has collaborated on experiments across the US, and in Germany, Japan, and the UK. Dr. Zarnstorff is a fellow of the American Physical Society and has served as a Division of Plasma Physics Distinguished Lecturer, and on the Executive Committee, Fellowship Committee, and Program Committee. He was a member of the NRC Plasma Science Committee, Burning Plasma Panel, and the Committee to Review the US ITER-Science Participation Planning. He has also served as a member of the DOE Fusion Energy Science Advisory Committee, and a number of sub-committees. Dr. Zarnstorff also served as Vice-Chair of the Council of the US Burning Plasma Organization, as Chair of the BPO International Collaboration Task Group, and on numerous advisory and review committees.

Index

A

ablate, 72, 122
ablated plasma, 135
ablation, 72, 77, 122
ablation front growth, 89
ablation pressure, 72
ablation surface, 78
ablative implosion, 73
activity level, 61
adaptive optics, 137
adiabat, 87
    adiabat-shaped 4-shock, 89
    high-adiabat, 88, 93
    low-adiabat, 88, 91
    low fuel adiabat, 89
    shaped implosions, 89
after heat, 49
alpha particles, 74
aluminum base, 142
Amplified Spontaneous Emission
   (ASE), 140
AP 1000, 30
armor, 149, 152
Au, 94
Au–Pd, 146
Au–Pd alloy coating, 143
Aurora Laser, 136

B

bandwidth, 132, 135
barriers, 48–49

basic nuclear installation, 225
beam engagement, 111
beam lines, 127
beam smoothing, 132
Beamlet, 85
beryllium, 210, 224
bio-shield, 216–217, 224
blanket, 125, 155, 159, 210
blanket designs, 151
blanket energy multiplication, 129
blanket gain B, 96–98, 108
boiling water reactors (BWR), 27
breed, 147
breeders, 17
breeding, 5
burning plasmas, 192

C

carbon fiber reinforced carbon
   (CFC), 212
carbon-free, 191
carbon-free energy, 190
CAREM, 41
cascade, 14
case-to-capsule ratio (CCR), 91
central hot spot, 72
central ignition, 76
central solenoid, 195, 207
centrifuge, 13, 21
chain reaction, 1–3, 6

Printed in the United States
By Bookmasters